SMART PRODUCT-SERVICE SYSTEMS

SMART PRODUCT-SERVICE SYSTEMS

PAI ZHENG

Department of Industrial and Systems Engineering,
The Hong Kong Polytechnic University,
Hung Hom, Kowloon, HKSAR

School of Mechanical and Aerospace Engineering,
Nanyang Technological University, Singapore

CHUN-HSIEN CHEN

School of Mechanical and Aerospace Engineering,
Nanyang Technological University, Singapore

ZUOXU WANG

School of Mechanical and Aerospace Engineering,
Nanyang Technological University, Singapore

Elsevier
Radarweg 29, PO Box 211, 1000 AE Amsterdam, Netherlands
The Boulevard, Langford Lane, Kidlington, Oxford OX5 1GB, United Kingdom
50 Hampshire Street, 5th Floor, Cambridge, MA 02139, United States

Copyright © 2021 Elsevier Inc. All rights reserved.

No part of this publication may be reproduced or transmitted in any form or by any means, electronic or mechanical, including photocopying, recording, or any information storage and retrieval system, without permission in writing from the publisher. Details on how to seek permission, further information about the Publisher's permissions policies and our arrangements with organizations such as the Copyright Clearance Center and the Copyright Licensing Agency, can be found at our website: www.elsevier.com/permissions.

This book and the individual contributions contained in it are protected under copyright by the Publisher (other than as may be noted herein).

Notices
Knowledge and best practice in this field are constantly changing. As new research and experience broaden our understanding, changes in research methods, professional practices, or medical treatment may become necessary.

Practitioners and researchers must always rely on their own experience and knowledge in evaluating and using any information, methods, compounds, or experiments described herein. In using such information or methods they should be mindful of their own safety and the safety of others, including parties for whom they have a professional responsibility.

To the fullest extent of the law, neither the Publisher nor the authors, contributors, or editors, assume any liability for any injury and/or damage to persons or property as a matter of products liability, negligence or otherwise, or from any use or operation of any methods, products, instructions, or ideas contained in the material herein.

Library of Congress Cataloging-in-Publication Data
A catalog record for this book is available from the Library of Congress

British Library Cataloguing-in-Publication Data
A catalogue record for this book is available from the British Library

ISBN: 978-0-323-85247-0

For information on all Elsevier publications visit our website at
https://www.elsevier.com/books-and-journals

Publisher: Matthew Deans
Acquisitions Editor: Brian Guerin
Editorial Project Manager: Emily Thomson
Production Project Manager: Prem Kumar Kaliamoorthi
Cover Designer: Matthew Limbert

Typeset by TNQ Technologies

Contents

Preface		*ix*
Acknowledgment		*xi*

1. Introduction   1
1.1	Transition toward digital servitization	1
1.2	Motivation and vision	3
1.3	Content organization	4
References		6

2. Evolvement of IT-driven product-service systems   9
2.1	A brief recap of product-service systems (PSS)	9
2.2	IT-driven PSS evolvement	15
2.3	Chapter summary	18
References		18

3. Fundamentals of smart product-service system   21
3.1	Basic notions	21
3.2	Technical aspect	27
3.3	Business aspect	28
3.4	Fundamentals of Smart PSS development	38
3.5	Chapter summary	46
References		46

4. Design entropy theory   53
4.1	Challenges of typical design methodologies	53
4.2	Design entropy theory	54
4.3	Self-adaptable design process of smart PSS	70
4.4	Information conversion map tool	72
4.5	Case study	75
4.6	Summary	81
References		81

v

vi   Contents

5. New IT-driven value co-creation mechanism   85
5.1 Value co-creation mechanism   87
5.2 Hybrid intelligence via crowd-sensing   98
5.3 Case study   109
5.4 Chapter summary   112
References   113

6. Graph-based context-aware product-service family configuration   117
6.1 Product-service family configuration in smart PSS   118
6.2 Graph-based product-service-context modeling   123
6.3 Requirement management based on the graph model   132
6.4 Solution configuration based on the hypergraph model   137
6.5 Case study   140
6.6 Chapter summary   147
References   148

7. Digital twin-enhanced product family design and optimization   151
7.1 Digital twin-enabled servitization   151
7.2 Trimodel-based generic framework   155
7.3 DT-enhanced product family design   160
7.4 DT-driven product family optimization   165
7.5 Case study featuring a context-aware DT system   170
7.6 Chapter summary   176
References   177

8. Engineering lifecycle implementations of smart product-service system   181
8.1 Design stage   183
8.2 Manufacturing stage   185
8.3 Distribution stage   187
8.4 Usage stage   189
8.5 End-of-life stage   191
8.6 Application scenarios   193
8.7 Chapter summary   196
References   196

9. Toward sustainable smart product-service systems   203
9.1 Two directions for promoting sustainability   203
9.2 Fundamentals of sustainable smart PSS (SSPSS)   205
9.3 Systematic framework for developing SSPSS   208

9.4 A four-phase PDCA procedure in the cyber space	210
9.5 Case study	218
9.6 Chapter summary	225
References	226

10. Conclusions and future perspectives	**229**
10.1 Conclusion	229
10.2 Future perspectives	232

Nomenclature	*235*
Index	*237*

Preface

Smart Product-Service System (Smart PSS), as a unique type of emerging PSS paradigm in the digital servitization era, is depicted as an IT-driven value cocreation business strategy by integrating smart, connected products and its generated digitalized and e-services into a single solution to meet individual customer needs in a sustainable manner. In this context, unlike conventional product or service, the design process starts from the very beginning of the lifecycle, Smart PSS development can be regarded as a closed-loop value generation process by taking both forward design and inverse design process into an overall consideration. Meanwhile, through high-fidelity modeling and massive data synchronization from heterogeneous sources, the digital twins of component, module, and products can be readily established to support product-service family design more cost-effectively. Moreover, enabled by the advanced ICT, users become actively engaged in the value cocreation process and their experience, i.e., user experience, drives the design innovation. Last but not least, AI plays a critical role for smart design decision-makings based on advanced machine intelligence (e.g., crowd sensing) or the human intelligence (e.g., crowd-sourcing), and novel approaches integrating both aspects should be explored and leveraged with context-awareness.

Given those facts, this book, as the first book in the Smart PSS field, aims to unveil the essence of it and to discuss the state-of-the-art enabling technologies to support engineering product-service development in to-day's smart, connected environment. Indeed, most of this book's contents are based on several valuable research projects been undertaken in Singapore and Hong Kong, including a comprehensive collection of novel findings and practical examples to demonstrate the unique aspects of Smart PSS, and to facilitate its industrial implementations cost-effectively. This book itself can be utilized as a reference book for university students on advanced courses, especially in the industrial engineering and business management field. The authors also hope the content of this book can attract ever increasing researchers' and practitioners' attention in this promising field, and welcome more open discussions and in-depth research and development to boost its valuable implementations in the digital servitization era.

Unlike conventional design processes that start from the very beginning of the lifecycle, design innovation can be regarded as a value generation process by considering the whole product-service lifecycle, including value cocreation at the early development stage, value implementation in the manufacturing process, and value recreation in the usage stage. It follows the Design for X principles with extended lifecycle and sustainability concerns, while conducted in a smart, connected environment by embracing state-of-the-art ICT and AI techniques.

Acknowledgment

The authors would like to give sincere thanks to the research team members, Ms. Jingchen Cong (Chapter 4), Mr. Kendrik Yan Hong Lim (Chapter 7), and Dr. Xinyu Li (Chapter 8) from the Design and Human Factors Lab at Nanyang Technological University, Singapore, for their contributions in accomplishing this book.

Meanwhile, the authors are also grateful for the funding support from National Research Foundation, Singapore, for the project entitled "Stakeholder-Oriented Innovative User-Centric Design and Solution Generation" (RCA-16/434), and National Nature Science Foundation of China for the project entitled "Research on key methodologies for user-oriented industrial smart product-service system" (No. 52005424).

Last but not least, the authors are also grateful to the colleagues at Elsevier, particularly Mr. Brian Guerin (Acquisitions Editor), Ms. Julie Luanco (Editorial Project Manager), and Prem Kumar Kaliamoorthi (Production Project Manager), for their warm support and painstaking efforts, which have ensured the smooth publication of this book.

CHAPTER 1

Introduction

Contents

1.1 Transition toward digital servitization	1
1.2 Motivation and vision	3
1.3 Content organization	4
References	6

The rapid development of information and communication technologies (ICT) has enabled the prevailing digital transformation (i.e., digitalization), where physical products can be readily digitized into the virtual space and maintain seamlessly interconnected (Zheng, Wang, Chen, & Khoo, 2019). Meanwhile, industries are ever-increasingly adopting service-oriented business models (i.e., servitization), to offer not only customized products but also tailored services as a solution bundle to meet individual customer needs (Green, Davies, & Ng, 2017). Such convergence of both digitalization and servitization, widely known as digital servitization, has triggered an emerging IT-driven business paradigm, Smart PSS. In this context, a large amount of low cost, high performance smart, connected products (SCPs) have been introduced and further leveraged as the tool and media to generate on-demand digitalized services. They together form the customized solution bundle, and ultimately delivered to the stakeholders in a smart, connected environment. To better understand the development of Smart PSS, as the prerequisite of this book, one should be aware of today's prevailing tendency toward digital servitization.

1.1 Transition toward digital servitization

Human beings are living in a modern digital era, of which numerous disruptive digital technologies, such as the Internet-of-Things (IoT), big data, artificial intelligence (AI), and edge-cloud computing have dramatically changed our ways of living. For instance, one can use the Uber app for taxi service and a robot vacuum cleaner for automatic room cleaning services. Meanwhile, for industrial companies, digital technologies enabled radical changes in products, services, innovation processes, business models,

Smart Product-Service Systems
ISBN 978-0-323-85247-0
https://doi.org/10.1016/B978-0-323-85247-0.00012-8

© 2021 Elsevier Inc.
All rights reserved.

and the fundamental nature of business activities in industrial ecosystems as well (Sklyar, Kowalkowski, Tronvoll, & Sörhammar, 2019; Sjodin, Parida, Jovanovic & Visjnic, 2020). For example, companies adopt "smart box" (container) for better logistic services (Zhang, Liu, Liu, & Li, 2016), and cloud manufacturing platforms for on-demand manufacturing resource allocation (Xu, 2012).

This depicts today's prevailing digital transformation, also known as digitalization, which refers to the use of digital technologies to offer new value-creating and revenue-generating opportunities (Parida, Sjödin, & Reim, 2019). Interestingly, digitalization typically goes "hand in hand with adopting a servitization strategy" (Parida, Sjodin, Lenka, & Wincent, 2015, p. 41), to boost the overall performance. Such convergence of digitalization and servitization is widely known as digital servitization, which denotes the utilization of digital tools for the transformational processes whereby a company shifts from a product-centric to a service-centric business model and logic (Tronvoll, Sklyar, Sörhammar, & Kowalkowski, 2020). Ever since the concept was first coined in 2015 (Lerch & Gotsch, 2015), many definitions and concepts have been brought up, of which the most widely accepted one was defined as "the transformation in processes, capabilities, and offerings within industrial firms and their associate ecosystems to progressively create, deliver, and capture increased service value arising from a broad range of enabling digital technologies" (Kohtamäki, Parida, Oghazi, Gebauer, & Baines, 2019; Parida et al., 2019).

In this context, digitalization is both the driver and enabler of servitization (Parida et al., 2015; Vendrell-Herrero, Bustinza, Parry, & Georgantzis, 2017), which may derive novel digital servitization innovation and business models among industrial companies (Rönnberg Sjödin, Parida, & Wincent, 2016). For instances, digital servitization enables active user participation, and refines the company's strategy for value proposition in a cocreation manner to meet customer's ever-evolving needs (Lenka, Parida, & Wincent, 2017). Meanwhile, product manufacturers adopt a digital servitization strategy to differentiate themselves with other competitors (Opresnik & Taisch, 2015) and to explore new channels of revenue streams (Parida et al., 2019).

However, digital servitization can be a double-edged sword, which creates both opportunities and challenges (i.e., digital servitization paradox) for industrial companies. For the good points, manufacturing companies are ever-increasingly investing in developing SCPs (Porter & Heppelmann, 2014) combined with AI capabilities to delivery smart functionalities. For instance, ABB offered enhanced digital service on their PLCs of YuMi robot to allow users to remote monitor and analyze its operation condition on site. Meanwhile, industrial firms, such as Siemens are also investing strategically in their Industrial IoT development (i.e., MindSphere), big data-analytics and digital twin services to make revenues. However, for the bad points, many companies are unable to cope with the speed of such transformation and fail to balance the cost and revenue of it, which is known as the digital servitization paradox. Certainly, digital technologies can effectively enhance quality and improve efficiency, however service costs are driven higher by the growing availability of more advanced solutions and capabilities (Sjödin et al., 2020), which request higher entry investment and maintenance costs (Porter & Heppelmann, 2015). Hence, many industrial firms remain hesitated or concerned about how to best address these issues properly. For instance, General Electric recently announced that their expenses on the Predix software platform would be cut by more than 25% (nearly $0.4 billion), revealing the difficulties involved. To overcome the challenges faced and also to explore the new opportunities, many recent works have been proposed, and the readers may refer to Paschou, Rapaccini, Adrodegari, & Saccani (2020) for more information.

Instead of investigating from a business oriented perspective, this book, entitled "Smart Product-Service System" as a typical paradigm of digital servitization, aims to address its fundamental issues from a technology-driven manner to boost industrial companies' smart solution development and implementations in a cost-effective manner.

1.2 Motivation and vision

With the arriving era of digital servitization, what fundamentally motivates the authors to write this book is to answer the following three questions:
- What are the unique aspects of Smart PSS compared with other existing PSS paradigms?
- How to ensure product-service (solution) development success in today's smart, connected environment?
- Why is it important to understand Smart PSS and conduct further research and development works on it?

The responses to the first question denotes the essence, the second one makes it valuable, and the third one reveals its significance, respectively. To some extent, these three questions are interrelated, however remain not yet well-addressed nor well-explained in most existing studies.

Meanwhile, another motivation comes from the fact that the book authors have been dedicated in Smart PSS field for years with some explorative research outcomes, and they are more than delighted to refine all of their materials contributed in the past and to offer a holistic and stepwise guide for Smart PSS development in both academia and industry.

Indeed, most of this book's contents are based on several valuable research projects undertaken in Singapore and Hong Kong, including a comprehensive collection of novel findings and practical examples to demonstrate the unique aspects of Smart PSS, and to facilitate its industrial implementations with cost-effectiveness. It is also suggested that the book can be utilized as a reference book for university students on advanced courses, especially in the business management or industrial engineering field. The authors hope the content of this book can attract ever increasing researchers' and practitioners' attention in the Smart PSS field, and welcome more open discussions and in–depth research and development to boost its valuable implementations in the digital servitization era.

1.3 Content organization

As the first published book in this field, we will offer a comprehensive introduction of Smart PSS, consisting of its core/emergence, definitions, uniqueness, challenges, development methodologies, lifecycle management, sustainability concerns, industrial cases, and potential future research and development directions. The main organization of this book is depicted in Fig. 1.1, including 10 chapters in total. Except for the overall introduction in Chapter 1, and the wrap-up and future highlights in Chapter 10, the remaining book chapters aim to answer the three key questions (i.e., What, How, and Why) raised accordingly. To distinguish Smart PSS with the other PSS paradigms (What), Chapter 2 describes the typical types of PSS in literature and the digital servitization evolvement of IT-driven PSS paradigms toward Smart PSS in today's smart, connected environment. Meanwhile, Chapter 3 further illustrates the fundamentals of Smart PSS and its development from both technical, economic, and sustainable concerns to present its uniqueness. To address the second question (How) holistically, Chapters 4−7 present the advanced development methodologies in correspondence with the key characteristics of Smart PSS. Chapter 4 introduces the design entropy theory for Smart PSS development by considering both design and reverse design in a closed-loop and self-adaptable manner. Chapter 5 investigates the hybrid value co-creation

Introduction 5

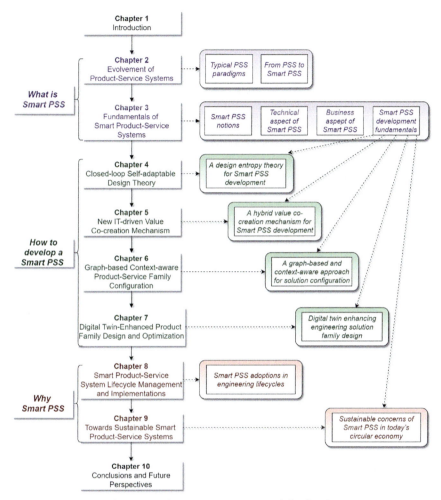

Figure 1.1 Content organization of this book.

mechanism of Smart PSS by integrating both human and machine intelligence as a whole, and further describes the core phases of value co-creation process. Chapter 6 presents the data-driven, graph-based, context-aware product-service family modeling and configuration processes and offers useful methodologies for enabling its solution recommendation/prediction. Chapter 7 reveals how the digital twin can enhance engineering solution family design and optimization process, by establishing a generic trimodel-based reference model and introducing the benchmarking, interacting, redesign and reconfiguration mechanisms, respectively. To explore the significance and value of Smart PSS (Why), Chapters 8 and 9 further discuss about its engineering lifecycle implementations with sustainable concerns. Chapter 8 summarizes the core implementation strategies and industrial cases of Smart PSS along different engineering lifecycle stages. Meanwhile, Chapter 9 discusses the sustainable concerns of Smart PSS in today's circular economy, and further propose a systematic framework with detailed procedures to guide its development toward sustainability based on the ISO14001:2015 standards.

References

Economics, I. J. P., Green, M. H., Davies, P., & Ng, I. C. L. (2017). Two strands of servitization: A thematic analysis of traditional and customer co-created servitization and future research directions. *International Journal of Production Economics, 192*, 40–53. https://doi.org/10.1016/j.ijpe.2017.01.009.

Kohtamäki, M., Parida, V., Oghazi, P., Gebauer, H., & Baines, T. (June 2019). Digital servitization business models in ecosystems: A theory of the firm. *Journal of Business Research, 104*, 380–392. https://doi.org/10.1016/j.jbusres.2019.06.027.

Lenka, S., Parida, V., & Wincent, J. (2017). Digitalization capabilities as enablers of value co-creation in servitizing firms. *Psychology and Marketing, 34*(1), 92–100. https://doi.org/10.1002/mar.

Lerch, C., & Gotsch, M. (2015). Digitalized product-service systems in manufacturing firms: A case study analysis. *Research-Technology Management, 58*(5), 45–52. https://doi.org/10.5437/08956308X5805357.

Opresnik, D., & Taisch, M. (2015). The value of big data in servitization. *International Journal of Production Economics, 165*, 174–184.

Parida, V., Sjödin, D. R., Lenka, S., & Wincent, J. (2015). Developing global service innovation capabilities: How global manufacturers address the challenges of market heterogeneity. *Research-Technology Management, 58*(5), 35–44. https://doi.org/10.5437/08956308x5805360.

Parida, V., Sjödin, D., & Reim, W. (2019). Reviewing literature on digitalization, business model innovation, and sustainable industry: Past achievements and future promises. *Sustainability, 11*(2). https://doi.org/10.3390/su11020391.

Paschou, T., Rapaccini, M., Adrodegari, F., & Saccani, N. (2020). Digital servitization in manufacturing: A systematic literature review and research agenda. *Industrial Marketing Management, 0*–1. https://doi.org/10.1016/j.indmarman.2020.02.012.

Porter, M. E., & Heppelmann, J. E. (2014). How smart, connected products are transforming competition. *Harvard Business Review, 92*(11), 64–88.

Porter, M. E., & Heppelmann, J. E. (2015). *Connected products are transforming companies how smart* (pp. 1–9). https://doi.org/10.1017/CBO9781107415324.004.

Rönnberg Sjödin, D., Parida, V., & Wincent, J. (2016). Value co-creation process of integrated product-services: Effect of role ambiguities and relational coping strategies. *Industrial Marketing Management, 56*, 108–119. https://doi.org/10.1016/j.indmarman.2016.03.013.

Sjödin, D., Parida, V., Kohtamäki, M., & Wincent, J. (January 2020). An agile co-creation process for digital servitization: A micro-service innovation approach. *Journal of Business Research, 112*, 478–491. https://doi.org/10.1016/j.jbusres.2020.01.009.

Sklyar, A., Kowalkowski, C., Tronvoll, B., & Sörhammar, D. (November 2019). Organizing for digital servitization: A service ecosystem perspective. *Journal of Business Research, 104*, 450–460. https://doi.org/10.1016/j.jbusres.2019.02.012.

Tronvoll, B., Sklyar, A., Sörhammar, D., & Kowalkowski, C. (January 2020). Transformational shifts through digital servitization. *Industrial Marketing Management, 89*, 293–305. https://doi.org/10.1016/j.indmarman.2020.02.005.

Vendrell-Herrero, F., Bustinza, O. F., Parry, G., & Georgantzis, N. (2017). Servitization, digitization and supply chain interdependency. *Industrial Marketing Management, 60*, 69–81. https://doi.org/10.1016/j.indmarman.2016.06.013.

Xu, X. (2012). From cloud computing to cloud manufacturing. *Robotics and Computer-Integrated Manufacturing, 28*(1), 75–86. https://doi.org/10.1016/J.RCIM.2011.07.002.

Zhang, Y., Liu, S., Liu, Y., & Li, R. (2016). Smart box-enabled product–service system for cloud logistics. *International Journal of Production Research, 54*(22), 6693–6706. https://doi.org/10.1080/00207543.2015.1134840.

Zheng, P., Wang, Z., Chen, C.-H., & Pheng Khoo, L. (April 2019). A survey of smart product-service systems: Key aspects, challenges and future perspectives. *Advanced Engineering Informatics, 42*, 100973. https://doi.org/10.1016/j.aei.2019.100973.

CHAPTER 2

Evolvement of IT-driven product-service systems

Contents

2.1 A brief recap of product-service systems (PSS)	9
2.2 IT-driven PSS evolvement	15
2.2.1 Internet-based PSS	16
2.2.2 IoT-enabled PSS	16
2.2.3 Smart PSS	17
2.3 Chapter summary	18
References	18

2.1 A brief recap of product-service systems (PSS)

In today's highly competitive market, industrial companies are gradually shifted from a product-oriented business model toward a service-dominant logic (Vargo & Lusch, 2004), by offering personalized products and services as a solution bundle to achieve higher user stickiness, profitability and sustainability. This paradigm is known as product-service systems (PSS) (Mont, 2002; Tukker & Tischner, 2006), which is a business strategy for value creation with lifecycle concerns, realized through cooperation and interaction among the different stakeholders (e.g., customers, OEMs, and suppliers) (Meier, Roy, & Seliger, 2010).

PSS was first coined in 1999 (Goedkoop, 1999) and defined as "a system composed of service and product element to provide values for relevant stakeholders." Ever since then, PSS has been widely investigated in both academia and industry as a promising approach to innovate product-based service offerings with competitiveness (Tukker, 2015). By adopting PSS, manufacturers' emphasis has progressively shifted to a *servitization* manner, by providing personalized products with value-added services (Parida, Sjödin, Wincent, & Kohtamäki, 2014). Hence, product becomes the tool and media to enable service innovation (Kuo & Wang, 2012), and the goal of PSS is to deliver not only a product (by ownership) but also its performance (e.g., pay-per-performance) and usage (e.g., pay-per-use, renting,

---

Smart Product-Service Systems
ISBN 978-0-323-85247-0
https://doi.org/10.1016/B978-0-323-85247-0.00011-6

© 2021 Elsevier Inc.
All rights reserved.

etc.) as a bundle (Marilungo, Papetti, Germani, & Peruzzini, 2017) to meet individual customer needs.

Meanwhile, to better address the specific issues of PSS, namely, its business models, service objects, sustainability concerns, and IT-driven perspectives, many different types of PSS paradigms have been introduced in literature. The authors summarized the definition of the first work of each category in Table 2.1 and briefly depicted below.

Business model-oriented PSS concerns the business strategies of a PSS paradigm the companies undertake. Based on the increasing degree of servitization (Baines et al., 2007), it can be classified into *product-oriented PSS*, *use-oriented PSS*, and *result-oriented PSS*, accordingly. Meanwhile, based on the form of servitization (e.g., performance, ownership, etc.) in each type, they can be further classified into eight types of business models with economic and environmental sustainability concerns (Tukker, 2004b), including (1) product-related services; (2) advice and consultancy in the *product-oriented PSS*; (3) product lease; (4) renting or sharing; (5) product pooling in the *use-oriented PSS*; (6) activity management/outsourcing; (7) pay per service unit; and (8) functional results in the *result-oriented PSS*, respectively.

Service object-oriented PSS emphasizes the target object the PSS adopted. The service object can be the human beings, i.e., customer/customization-oriented PSS, which mainly investigates the relationship between the end users of the PSS and its designers to offer better and customized solution design to the stakeholders (e.g., customer satisfaction of the iPad). The service object can also be the specific technical product, i.e., technical PSS focuses the life cycle—oriented services to enhance the performance/quality of the technical products (e.g., maintenance services of car engine). Moreover, the service object can be the industrial applications, i.e., industrial PSS, which is characterized by the integrated and mutually determined planning, development, provision, and use of products and services within the manufacturing plant.

Sustainability-oriented PSS mainly aims to address the sustainability issues of the PSS deployment. Although the introduction of PSS is to enhance the sustainability of product-oriented business model, indeed the adoption of PSS may not by nature ensure or improve the social, economic, and environmental sustainability. Hence, the concept of sustainable PSS, smart sustainable PSS, and smart-circular PSS have been brought up recently to facilitate its development and assessment toward better sustainability in today's circular economy.

Table 2.1 Classification of typical PSS paradigms.

Category	Name	Definition	References
Business types	Product-oriented PSS	"Product-oriented PSS: promoting/selling the product in a traditional manner, while including in the original act of sale additional services such as, after sales service to guarantee functionality and durability of the product owned by the customer (maintenance, repair, re-use and recycling, and helping customers optimise the application of a product through training and consulting). The company is motivated to introduce a PSS to minimise costs for a long lasting, well-functioning product and to design products to take account of product end-of-life (re-useable/ easily replaceable/ recyclable parts)."	Baines et al. (2007)
	Use-oriented PSS	"Use-oriented PSS: selling the use or availability of a product which is not owned by the customer (e.g., leasing, sharing). In this case the company is motivated to create a PSS to maximise the use of the product needed to meet demand and to extend the life of the product and materials used to produce it."	

Continued

12 Smart Product-Service Systems

Table 2.1 Classification of typical PSS paradigms.—cont'd

Category	Name	Definition	References
	Result-oriented PSS	"Result-oriented PSS; selling a result or capability instead of a product (e.g., web information replacing directories, selling laundered clothes instead of a washing machine). Companies offer a customised mix of services where the producer maintains ownership of the product and the customer pays only for the provision of agreed results."	
Service object	Customer/Customization-oriented PSS	"Customer-oriented model includes layers about customers (and customer groups), customer barriers and solutions."	Schmidt, Malaschewski, Fluhr, and Mörtl (2015)
	Technical PSS	"A technical PSS consists of a physical product core enhanced and individualized by a mainly non-physical technical service shell that is realized throughout the entire life cycle of the physical product core."	Aurich, Fuchs, and Wagenknecht (2006)
	Industrial PSS	"Industrial PSS are made up of a complex physical product core dynamically enhanced along its life cycle by mainly non-physical services."	Aurich, Schweitzer, and Fuchs (2007)

Evolvement of IT-driven product-service systems **13**

Table 2.1 Classification of typical PSS paradigms.—cont'd

Category	Name	Definition	References
Sustainability	Sustainable PSS	"When a PSS assists re-orient current unsustainable trends in production and in consumption practices, it is usually referred to as a Sustainable or Eco-efficient PSS."	Manzini and Vezolli (2003)
	Smart-circular PSS	"As smart enablers are foundational building blocks of smart products, several technological capabilities are combined for the operationalisation of smart-circular PSS."	Alcayaga, Wiener, and Hansen (2019)
	Sustainable Smart PSS	"The concept of Sustainable Smart PSS can be regarded as the trinary intersection of sustainable strategy, smart technology, and PSS."	Li, Wang, Chen, and Zheng (2021)
Information technology	Internet-based/web-based PSS	"Internet, web enabled and wireless technologies help customers to have more and more options in front of them at an instant of a click. Their access to more suppliers and products through Internet unlimits their power of purchase, and empower them to ask better products at a minimum cost with extended service additions."	Lee (2003)

Continued

14 Smart Product-Service Systems

Table 2.1 Classification of typical PSS paradigms.—cont'd

Category	Name	Definition	References
	IoT-enabled PSS	"The concept of the Product-service system (PSS) has been increasingly used as a new business model in implementing advanced technologies including RFID."	Baines et al. (2010)
	Smart PSS	"Smart products and its generated e-services into a single solution by embracing disruptive ICT."	Valencia, Mugge, Schoormans, and Schifferstein (2015)
	Digitalized PSS	"Manufacturers at this stage not only provide complex PSS to their customers, but also incorporate ICT solutions as a novel component in the product-service bundle, creating intelligent, independent operating systems that deliver the highest level of availability possible and optimize operations while reducing resource inputs."	Lerch and Gotsch (2015)
	Cyber-Physical PSS	"A combination of CPS technology with PSS business models."	Wiesner and Thoben (2017)

IT-driven PSS represents the embracement of disruptive IT to facilitate the PSS development process, which emphasizes the influential impact of IT innovations on the PSS and its development. During the past two decades, we have witnessed three digital waves, including internet, mobile internet, and IoT, with almost 20 billion of end devices connected to the internet to date. Moreover, we are also embracing the upcoming fourth digital wave of digital twin (DT), cyber-physical system (CPS), and industrial artificial intelligence. They all have changed the way PSS develops and operates dramatically.

The content of this book mainly deals with the most advanced IT-driven PSS paradigm, i.e., Smart PSS, and as a prerequisite, the authors will further discuss the evolvement of those IT-driven paradigms in the following section.

2.2 IT-driven PSS evolvement

To the authors' best knowledge, the IT-driven PSS evolvement can be divided into three phases ever since PSS was first coined in 1999 (Goedkoop, 1999), beginning with internet-based PSS (1999−) (conventional PSS) (Lee, 2003) and IoT-enabled PSS (2010−) (Michael et al., 2010), to today's *Smart PSS* (Valencia et al., 2015), as shown in Fig. 2.1. The vertical axis represents the degree of digital servitization of each phase, while the horizontal axis stands for the timeline of the IT innovation. The key technologies and significant academic milestones of each phase are listed in Fig. 2.1 and elaborated below.

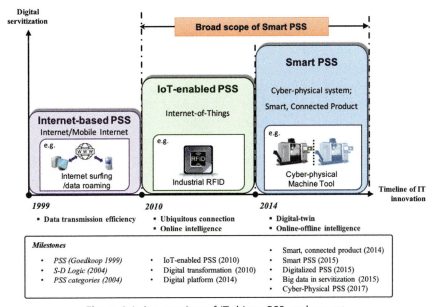

Figure 2.1 An overview of IT-driven PSS evolvement.

2.2.1 Internet-based PSS

Internet-based PSS benefits from the wide usage of internet/mobile internet ever since the emergence of PSS. It follows the basic principles of PSS as a product and e-service bundle in the internet environment, and the major concern still lies in the efficient delivery of data/information (e.g., 2G/3G/4G) with little digitalization consideration. In this phase, the first publication can be traced back to 1999, when web-based computer-aided maintenance was proposed by Lee and AbuAli (Lee & AbuAli, 2011). In 2003, Lee further introduced the first concept of "smart product and service systems" (Lee, 2003) by leveraging the internet to provide e-maintenance services. Although their understanding and definition of Smart PSS largely differs from today's one, it is interesting to mark its novelty and contribution to future works. Meanwhile, other key milestones such as service-dominant logic (S-D Logic) (Vargo & Lusch, 2004), and eight types of PSS business models (Tukker, 2004a) also set the foundation of servitization in this phase.

2.2.2 IoT-enabled PSS

In this phase, IoT (viz. WSN and RFID) is largely accepted as the enabling technology with the highest economic impact for PSS development (Marilungo et al., 2017). It has been widely adopted in manufacturing enterprises (e.g., automobile, medical devices, etc.) to enhance product value and create novel services for stakeholders, segmented into a B2C or B2B market (Takenaka, Yamamoto, Fukuda, Kimura, & Ueda, 2016).

IoT-enabled PSS is triggered by the prevailing adoption of IoT (Kevin, 2009), as the key technology to drive the digital transformation (Yoo, Henfridsson, & Lyytinen, 2010). IoT-enabled PSS (Michael et al., 2010) was first introduced by exploiting RFID devices in the industrial environment, where RFID readers/tags (product) and their embedded information are connected to the internet for production maintenance purposes (digital service). Moreover, platform approach was adopted as an organizational business perspective to leverage the value of IT-enabled interactions and digital technologies based on modularity (Chiu, Chu, & Kuo, 2019; Thomas, Autio, & Gann, 2014). Hence, in this phase, the major concern exceeds the efficient delivery of data/information as in the internet-based PSS, but to take both online information processing capability and modular service composition (cloud services) into consideration for digitization and servitization.

2.2.3 Smart PSS

Driven by the rapid development of information and communication technologies (ICT), low cost, high performance embedded components are ever increasingly introduced by manufacturers so as to embrace a promising market of information densely product, i.e., SCP (Porter & Heppelmann, 2014). Apart from the mechanism of IoT by providing a ubiquitous connectivity at a low cost for information transmitting, SCP changes the way how value is created by embedding IT (i.e., smart and connectivity components) into the product itself (Porter & Heppelmann, 2015a). Owing to its capabilities of collect, process, produce information, and somehow to "think by itself" (Rijsdijk & Hultink, 2009), SCPs can be readily digitized into the virtual space and remain interconnected with the physical ones, which represents the prevailing digital transformation (Loonam, Eaves, Kumar, & Parry, 2018). Also, they can serve as the media and tools for generating value-added services as a solution bundle (Lerch & Gotsch, 2015) (Djellal et al., 2018).

Smart PSS was coined in 2014 (Valencia et al., 2015) and enabled by the wide adoption of SCPs (Porter & Heppelmann, 2015b, p. 9). In this context, huge amount of data is generated by SCPs from multiple sources across ad hoc connections, and ultimately transmitted to stakeholders through various analytic tools and business intelligence (Rymaszewska, Helo, & Gunasekaran, 2017). An informatic-based approach serves as the key for value creation by providing the ability to identify user behavior patterns or latent needs (Lim, Kim, Heo, & Kim, 2015). Meanwhile, it is also motivated by the digitalization of products and services enabled by the CPS (Wiesner & Thoben, 2017) and Big Data technologies (Opresnik & Taisch, 2015), where the DT (Grieves, 2014) established between components allow autonomous interaction and further optimization of them, which represents the cutting-edge digital servitization paradigm (Lerch & Gotsch, 2015).

Hence, in this phase, despite the ubiquitous connectivity and online intelligence enabled by the IoT-enabled PSS, both cyberspace (online) and physical space (offline) should also be considered to realize Smart PSS. Furthermore, the degree of "smartness" can be further divided into five levels (i.e., smart, connectivity (1), smart analytics (2), DT (3), cognition (4), and autonomous (5)) according to the 5C architecture (Lee, Ardakani, Yang, & Bagheri, 2015; Zheng, Lin, Chen, & Xu, 2018).

Nevertheless, to the authors' knowledge, some articles argue that Smart PSS starts from 2010 onwards (Alcayaga, Hansen, & Wiener, 2019; Luz Martín-Peña, Díaz-Garrido, & Sánchez-López, 2018). It is partially agreed since the adoption of IoT marks the infancy stage of digital transformation (Loonam et al., 2018), and serves as the prerequisite for Smart PSS (i.e., ubiquitous connectivity). However, this book respects the truth when the terminology was first coined in 2014 and widely accepted thereafter, with many new technologies exploited other than IoT. Hence, to avoid any bias, the authors treat the IoT-enabled PSS as an extended scope of Smart PSS, as shown in Fig. 2.1.

2.3 Chapter summary

This chapter provided a brief recap of the typical PSS paradigms, and outlined the IT-driven PSS evolvement in history. It can be found that among the various categories of PSS paradigms, IT-driven PSS plays a significant role by leveraging the advanced digital technologies to enable its development and implementations. Smart PSS, as the state-of-the-art IT-driven PSS paradigm, is empowered by the current third wave and the upcoming fourth of IT innovation for value generation, of which SCPs act as the tool and media, together with the advanced IT infrastructure to generate various services as the solution bundle to meet individual customer needs. All the content described in this chapter can offer the readers a better understanding of the motivation and scope of this book.

References

Alcayaga, A., Hansen, E. G., & Wiener, M. (2019). Towards a framework of smart-circular systems: An integrative literature review. *Journal of Cleaner Production, 221*, 622–634. https://doi.org/10.1016/j.jclepro.2019.02.085.

Aurich, J. C., Fuchs, C., & Wagenknecht, C. (2006). Life cycle oriented design of technical Product-Service Systems. *Journal of Cleaner Production, 14*(17), 1480–1494.

Aurich, J. C., Schweitzer, E., & Fuchs, C. (2007). Life cycle management of industrial product-service systems. In *Advances in life cycle engineering for sustainable manufacturing businesses* (pp. 171–176). Springer.

Baines, T. (2010). Manufacturing operations strategy –3rd edition, by A. Hill and T. Hill. *International Journal of Production Research, 48*(12), 3709. https://doi.org/10.1080/00207540903290159.

Baines, T. S., Lightfoot, H. W., Evans, S., Neely, A., Greenough, R., Peppard, J., ... Wilson, H. (2007). State-of-the-art in product-service systems. *Proceedings of the Institution of Mechanical Engineers, Part B: Journal of Engineering Manufacture, 221*(10), 1543–1552. https://doi.org/10.1243/09544054JEM858.

Chiu, M., Chu, C., & Kuo, T. C. (2019). Product service system transition method: Building firm's core competence of enterprise. *International Journal of Production Research, 0*(0), 1−21. https://doi.org/10.1080/00207543.2019.1566670.

Djellal, F., Gallouj, F., Saccani, N., Adrodegari, F., Perona, M., & Paschou, T. (2018). Digital servitization in manufacturing as a new stream of research: A review and a further researc. *A Research Agenda for Service Innovation,* 148−165. https://doi.org/10.4337/9781786433459.00012.

Goedkoop, M. J., Van Halen, C. J. G., Te Riele, H. R. M., & Rommens, P. J. M. (1999). Product service systems, ecological and economic basics. *Report for Dutch Ministries of environment (VROM) and economic affairs (EZ), 36*(1), 1−122.

Grieves, M. (2014). *Digital twin: Manufacturing excellence through virtual factory replication* (pp. 1−7). Retrieved from: http://innovate.fit.edu/plm/documents/doc_mgr/912/1411.0_Digital_Twin_White_Paper_Dr_Grieves.pdf.

Kevin, A. (2009). That 'internet of things' thing. *RFID Journal, 22*(7), 97−114.

Kuo, T. C., & Wang, M. L. (2012). The optimisation of maintenance service levels to support the product service system. *International Journal of Production Research, 50*(23), 6691−6708. https://doi.org/10.1080/00207543.2011.616916.

Lee, J. (2003). Smart products and service systems for e-business transformation. *International Journal of Technology Management, 26*(1), 45. https://doi.org/10.1504/ijtm.2003.003143.

Lee, J., & AbuAli, M. (2011). Innovative product advanced service systems (I-PASS): Methodology, tools, and applications for dominant service design. *International Journal of Advanced Manufacturing Technology, 52*(9−12), 1161−1173.

Lee, J., Ardakani, H. D., Yang, S., & Bagheri, B. (2015). Industrial big data analytics and cyber-physical systems for future maintenance & service innovation. *Procedia CIRP, 38,* 3−7. https://doi.org/10.1016/j.procir.2015.08.026.

Lerch, C., & Gotsch, M. (2015). Digitalized product-service systems in manufacturing firms: A case study analysis. *Research-Technology Management, 58*(5), 45−52. https://doi.org/10.5437/08956308X5805357.

Li, X., Wang, Z., Chen, C.-H., & Zheng, P. (2021). A data-driven reversible framework for achieving sustainable smart product-service systems. *Journal of Cleaner Production, 279,* 123618. https://doi.org/10.1016/j.jclepro.2020.123618.

Lim, C.-H., Kim, M.-J., Heo, J.-Y., & Kim, K.-J. (2018). Design of informatics-based services in manufacturing industries: case studies using large vehicle-related databases. *Journal of Intelligent Manufacturing, 29*(3), 497−508. https://doi.org/10.1007/s10845-015-1123-8.

Loonam, J., Eaves, S., Kumar, V., & Parry, G. (2018). Towards digital transformation: Lessons learned from traditional organizations. *Strategic Change, 27*(2), 101−109. https://doi.org/10.1002/jsc.2185.

Luz Martín-Peña, M., Díaz-Garrido, E., & Sánchez-López, J. M. (2018). The digitalization and servitization of manufacturing: A review on digital business models. *Strategic Change, 27*(2), 91−99. https://doi.org/10.1002/jsc.2184.

Manzini, E., & Vezzoli, C. (2003). A strategic design approach to develop sustainable product service systems: examples taken from the `environmentally friendly innovation'Italian prize. *Journal of Cleaner Production, 11*(8), 851−857.

Marilungo, E., Papetti, A., Germani, M., & Peruzzini, M. (2017). From PSS to CPS design: A real industrial use case toward industry 4.0. *Procedia CIRP, 64,* 357−362. https://doi.org/10.1016/j.procir.2017.03.007.

Meier, H., Roy, R., & Seliger, G. (2010). Industrial product-service systems-IPS2. *CIRP Annals - Manufacturing Technology, 59*(2), 607−627. https://doi.org/10.1016/j.cirp.2010.05.004.

Michael, K., Roussos, G., Huang, G. Q., Chattopadhyay, A., Gadh, R., Prabhu, B. S., & Chu, P. (2010). Planetary-scale RFID Services in an age of uberveillance. *Proceedings of the IEEE, 98*(9), 1663−1671. https://doi.org/10.1109/JPROC.2010.2050850.

Mont, O. (2002). Clarifying the concept of product−service system. *Journal of Cleaner Production, 10*(3), 237−245. https://doi.org/10.1016/S0959-6526(01)00039-7.

Opresnik, D., & Taisch, M. (2015). The value of big data in servitization. *International Journal of Production Economics, 165*, 174–184.

Parida, V., Sjödin, D. R., Wincent, J., & Kohtamäki, M. (2014). Mastering the transition to product-service provision: Insights into business models, Learning activities, and capabilities. *Research Technology Management, 57*(3), 44–52. https://doi.org/10.5437/08956308X5703227.

Porter, M. E., & Heppelmann, J. E. (November 2014). How smart, connected product are transforming competition. *Harvard Business Review*, 64–89. https://doi.org/10.1017/CBO9781107415324.004.

Porter, M. E., & Heppelmann, J. E. (2015a). Connected products are transforming companies how smart. *Connected Products Are Transforming Companies, 1–9*. https://doi.org/10.1017/CBO9781107415324.004.

Porter, M. E., & Heppelmann, J. E. (2015b). *How smart, connected products are transforming companies* (p. 9).

Rijsdijk, S. A., & Hultink, E. J. (2009). How today's consumers perceive tomorrow's smart products. *Journal of Product Innovation Management, 26*(1), 24–42. https://doi.org/10.1111/j.1540-5885.2009.00332.x.

Rymaszewska, A., Helo, P., & Gunasekaran, A. (February 2017). IoT powered servitization of manufacturing — an exploratory case study. *International Journal of Production Economics, 192*, 92–105. https://doi.org/10.1016/j.ijpe.2017.02.016.

Schmidt, D. M., Malaschewski, O., Fluhr, D., & Mörtl, M. (2015). Customer-oriented Framework for Product-service Systems. *Procedia CIRP, 30*, 287–292. https://doi.org/10.1016/j.procir.2015.02.106.

Takenaka, T., Yamamoto, Y., Fukuda, K., Kimura, A., & Ueda, K. (2016). Enhancing products and services using smart appliance networks. *CIRP Annals - Manufacturing Technology, 65*(1), 397–400. https://doi.org/10.1016/j.cirp.2016.04.062.

Thomas, L. D. W., Autio, E., & Gann, D. M. (2014). Architectural leverage: Putting platforms in context. *Academy of Management Perspectives, 28*(2), 198–219.

Tukker, A. (2004a). Eight types of product-service system: Eight ways to sustainability? Experiences from suspronet. *Business Strategy and the Environment, 13*(4), 246–260. https://doi.org/10.1002/bse.414.

Tukker, A. (2004b). Eight types of product —service system: Eight ways to sustainability? *Experiences from suspronet, 260*, 246–260.

Tukker, A. (2015). Product services for a resource-efficient and circular economy - a review. *Journal of Cleaner Production, 97*, 76–91. https://doi.org/10.1016/j.jclepro.2013.11.049.

Tukker, A., & Tischner, U. (2006). Product-services as a research field: Past, present and future. Reflections from a decade of research. *Journal of Cleaner Production, 14*(17), 1552–1556. https://doi.org/10.1016/j.jclepro.2006.01.022.

Valencia, A., Mugge, R., Schoormans, J. P. L., & Schifferstein, H. N. J. (2015). The design of smart product-service systems (PSSs): An exploration of design characteristics. *International Journal of Design, 9*(1), 13–28. https://doi.org/10.1016/j.procir.2016.04.078.

Vargo, S. L., & Lusch, R. F. (2004). Evolving to a new dominant logic for marketing. *Journal of Marketing, 68*(1), 1–17. https://doi.org/10.1509/jmkg.68.1.1.24036.

Wiesner, S., & Thoben, K.-D. (2017). Cyber-physical product-service systems. In S. Biffl, A. Lüder, & D. Gerhard (Eds.), *Multi-disciplinary engineering for cyber-physical production systems* (pp. 63–88). Cham: Springer International Publishing.

Yoo, Y., Henfridsson, O., & Lyytinen, K. (2010). The new organizing logic of digital innovation: An agenda for information systems research. *Information Systems Research, 21*(4), 724–735. https://doi.org/10.1287/isre.1100.0322.

Zheng, P., Lin, Y., Chen, C.-H., & Xu, X. (2018). Smart, connected open architecture product: An IT-driven co-creation paradigm with lifecycle personalization concerns. *International Journal of Production Research, 0*(0), 1–14. https://doi.org/10.1080/00207543.2018.1530475.

CHAPTER 3

Fundamentals of smart product-service system

Contents

3.1 Basic notions	21
3.1.1 Definition and scope	22
3.1.2 Clarification of concepts	26
3.2 Technical aspect	27
3.2.1 Enabling digital technologies	27
3.2.2 Implementation architecture	28
3.3 Business aspect	28
3.3.1 Business model	28
3.3.2 Digital platform	32
3.3.3 Value co-creation	32
3.4 Fundamentals of Smart PSS development	38
3.4.1 Two ways of development process	38
3.4.2 Three key characteristics	40
3.4.3 Sustainability concerns	44
3.5 Chapter summary	46
References	46

The third wave of IT innovation has triggered a promising market of information-densely products, i.e., smart, connected products (SCPs), where IT are embedded in the physical products themselves to develop on-demand services in a software-defined manner. Both the SCPs and their generated e-services and digitalized services, as a smart product-service system (Smart PSS), enable today's prevailing business transition toward digital servitization. To better understand the essence of Smart PSS, this chapter will provide a fundamental introduction of its basic notions, key aspects, and development issues.

3.1 Basic notions

Ever since the term Smart PSS was coined in 2014, many relevant concepts and definitions have been proposed. By summarizing those shared fundamentals, this section provides a unified definition and scope of Smart PSS and clarifies some basic concepts for better readability.

Smart Product-Service Systems
ISBN 978-0-323-85247-0
https://doi.org/10.1016/B978-0-323-85247-0.00007-4

© 2021 Elsevier Inc.
All rights reserved.

3.1.1 Definition and scope

Among existing literature, the major definitions and key elements of Smart PSS are listed in Table 3.1, including both system level and ecosystem level (system-of-systems). One can find that, although lacking a unified definition, it is widely accepted that SCP and its generated services, as the solution bundle, are the fundamental composition of Smart PSS, and it undertakes an IT-driven value co-creation manner to fulfill customer needs. Meanwhile, similar terminologies are also summarized, as shown in Table 3.2, including cyber-physical PSS (Wiesner & Thoben, 2017) and digitalized PSS (Lerch & Gotsch, 2015). Both share the same fundamentals as Smart PSS, i.e., digitalization and servitization enabled by the advanced IT infrastructure and digital technologies.

Hence, to make the definition consistent throughout this book, and to include the digitalization of products and services (e.g., digital twin) as the broad scope of Smart PSS depicted in Chapter 2, the authors follow the definition of our previous work that:

> Smart PSS is an IT-driven value co-creation business strategy consisting of various stakeholders as the players, intelligent systems as the infrastructure, smart, connected products as the media and tools, and their generated services (i.e. e-services and digitalized services) as the key values delivered that continuously strives to meet individual customer needs in a sustainable manner.

According to this definition, the scope of Smart PSS can be further classified into three levels, namely: *product-service level*, *system level*, and *system-of-systems (ecosystem) level*, as shown in Fig. 3.1. The first two represent the widely accepted scope of Smart PSS or the narrow scope of it, while the third one denotes the broad scope.

Product-service level refers to a single solution bundle consisting of SCP, its generated services intend to meet individual user needs. *System level* expands the scope from the functionality of a discrete solution to the systematic integration of a family of solutions, the involvement of various stakeholders (e.g., users, service providers) and platform-based intelligent systems to optimize the overall performance. *System-of-systems level* further extends the industry boundaries to a set of associated Smart PSSs as well as the related external auxiliary systems (e.g., data management system) to make the greatest impact on the total ecosystems, such as the smart transportation system and smart building (Porter & Heppelmann, 2014).

It should be claimed that the content of this book mainly focuses on the narrowed scope of Smart PSS for its solution and system development, while the interactions between various associated Smart PSSs in an ecosystem is beyond our discussion.

Fundamentals of smart product-service system 23

Table 3.1 Typical definitions of smart PSS.

Smart PSS related terms	Reference	Definitions	Key elements
Smart PSS	Valencia Cardona, Mugge, Schoormans, and Schifferstein (2014), Valencia, Mugge, Schoormans, and Schifferstein (2015)	"Smart products and its generated e-services into a single solution by embracing disruptive ICT."	Organization-wise: stakeholder involvement. Design-wise: complex market offerings, ICT technologies, e-services, interactions between the Smart PSS and end-users; context, and a life-long development issue.
	Kuhlenkötter, Wilkens et al. (2017)	"A digital-based ecosystem of value creation characterized by high complexity, dynamics and interconnectedness among stakeholders."	/
	Zheng, Lin, Chen, and Xu (2018a), Zheng, Wang, and Chen (2019), Zheng, Chen, and Shang (2019)	"An IT-driven value co-creation business strategy consisting of various stakeholders as the players, intelligent systems as the infrastructure, smart, connected products as the media and tools, and their generated services as the key values delivered that continuously strives to meet individual customer needs in a sustainable manner."	SCPs and generated e-services Stakeholders Intelligent system (smart environment)

Continued

Table 3.1 Typical definitions of smart PSS.—cont'd

Smart PSS related terms	Reference	Definitions	Key elements
	Liu, Ming, Song, Qiu, and Qu (2018)	"A platform service ecosystem, in which platform is made up of smart products and smart services, while multiple service systems constitute a service ecosystem."	Service platform, SCPs, and smart services
Smart industrial product-service system-of-systems	Liu and Ming (2019)	"A consolidation of resources on people, intelligent machines and devices, intelligent objects, smart services including primary e-services and digital services, infrastructures and networks to fulfill customer needs as problems to be solved."	/
Smart product-service ecosystem	Zheng, Ming, Wang, Yin, and Zhang (2017)	"An ICT based dynamic ecological Smart PSS network, which integrates customers, smart product service systems, smart service platform and product service suppliers for value co-creation and customer experience improvement, by means of smart interaction, mutual cooperation, resource sharing and optimal configuration."	From the perspective of ecosystem: (1) connection, (2) business model, (3) relationship, (4) user experience, (5) technologies, and (6) marketing

Fundamentals of smart product-service system 25

Table 3.2 Other similar PSS terminologies.

Other PSS terms	Reference	Definitions
Cyber-physical PSS	Wiesner and Thoben (2017)	"Manufacturers at this stage not only provide complex PSS to their customers, but also incorporate ICT solutions as a novel component in the product-service bundle, creating intelligent, independent operating systems that deliver the highest level of availability possible and optimize operations while reducing resource inputs."
Digitalized PSS	Lerch and Gotsch (2015)	"A combination of CPS technology with PSS business models."

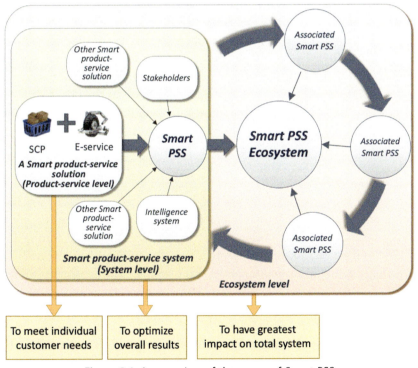

Figure 3.1 An overview of the scope of Smart PSS.

3.1.2 Clarification of concepts

This section aims to clarify some confusing concepts of Smart PSS for better understanding.

Digital twin (DT) versus *cyber-physical system (CPS)*. DT is defined as a dynamic virtual representation (i.e., digital mock-up) of what has been produced for product life-cycle management (PLM) in a data-driven manner. According to Grieve's study (Grieves, 2014, pp. 1—7), it mainly contains three parts: "(a) physical products in Real Space, (b) virtual products in Virtual Space, and (c) the connections of data and information that tie the virtual and real products together." CPS, on the other hand, stands for the "embedded computers and networks monitor and control the physical processes, usually with feedback loops where physical processes affect computations and vice versa" (Lee, 2008, pp. 363—369). Hence, CPS provides the high performance IT environment for the realization of high-fidelity simulation, monitoring, and control of DT.

Digitalized services versus *e-services*. Digitalized services refer to the digital forms of services that reflect the functionality or status of SCPs themselves, for instance, the remote monitoring and control of the aircraft engine based on its DT. However, e-services here stand for those digitalized add-on services that have little dependence with the functionality or status of SCPs. For instance, e-book or Uber as an e-service is independent with the SCPs (e.g., iPad, personal computer, etc.) utilized. In this book, the authors mainly address the prior one, defined as the narrowed scope of Smart PSS in Chapter 2.

Industrial Smart PSS versus *customer-oriented PSS*. The two forms of Smart PSS denote the business model the company is adopted internally and externally, respectively. The prior one emphasizes the smart solution development and implementation, as the business activities occurred inside the company and/or with other vendors, such as smart manufacturing and logistics. However, customer-oriented PSS, on the other hand, focuses on the value capture, proposition, and delivery of the end-users or customers via the smart solutions developed. Those business activities can include smart home and living applications. This book aims to depict the general principles lying behind for the companies to follow when developing their own smart solutions, no matter in a customer-oriented Smart PSS (e.g., examples in Chapter 4—6) or industrial one (e.g., examples in Chapter 7 and 8).

3.2 Technical aspect

The enabling digital technologies and system implementation architecture of Smart PSS are described in this section.

3.2.1 Enabling digital technologies

From technical aspect, to realize Smart PSS, the enabling digital technologies can be classified into three categories, i.e., *networking capability, intelligence capability,* and *analytic capability* (Ardolino et al., 2018; Ardolino, Saccani, Gaiardelli, & Rapaccini, 2016; Lenka, Parida, & Wincent, 2017; Parida, Sjödin, Lenka, & Wincent, 2015), and each prior category serves as the foundation for the latter one, as shown in Fig. 3.2.

Networking capability stands for the ability to connect various SCPs through wired and wireless communication networks. IoT serves as the fundamental technology to enable ubiquitous connectivity (connectedness), so as to enable manufacturers or service providers to collect data and offer product-related services based on the product operation data (Brehm & Klein, 2017; Rymaszewska, Helo, & Gunasekaran, 2017). Wireless communications can be achieved through radio-frequency identification (RFID), Bluetooth, global positioning system (GPS), wireless sensor network (WSN), etc.

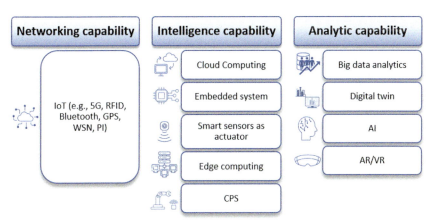

Figure 3.2 Enabling digital technologies of Smart PSS.

Intelligence capability denotes the ability to configure hardware components to sense and capture information with less human intervention into a so-called CPS. It is enabled either by the technologies to upgrade hardware components with smart subcomponents, such as embedded system, smart sensors, and actuators, or the computing technologies to collect and process massive information of the product-sensed data or user-generated data, such as cloud computing (Ardolino et al., 2016) and edge computing (Zheng & Chen, 2018).

Analytic capability represents the ability to transform the data/information available at hand into valuable insights and actionable directives for the company, which is enabled by the cutting-edge DT, AI, and Big Data analytics (Maglio & Lim, 2016).

All of them set the technical foundation to realize the Smart PSS, and can be further deployed in various SCPs for specific purposes, such as remote operation access by utilizing the augmented reality devices (Mourtzis, Zogopoulos, & Vlachou, 2017), or DT-enabled machine tool monitoring system (Tao et al., 2017).

3.2.2 Implementation architecture

Empowered by the essential digital technologies, implementation architecture of Smart PSS is of paramount importance (Kuhlenkötter, Bender, et al., 2017). Table 3.3 summarizes the typical system architectures of Smart PSS in literature, all of which follow the data-information-knowledge-wisdom (DIKW) model (Ardolino et al., 2018), to construct a reliable system for Smart PSS development.

3.3 Business aspect

This section outlines the key business models in Smart PSS, types of digital platforms, and the value co-creation process.

3.3.1 Business model

Smart PSS represents the advanced digital business model, i.e., digital servitization (Coreynen, Matthyssens, & Van Bockhaven, 2017; Vendrell-Herrero, Bustinza, Parry, & Georgantzis, 2017). Nevertheless, it follows the typical classifications defined in Tukker's study (Tukker, 2004), including *product-oriented services, use-oriented services,* and *result-oriented services,* with respective examples shown in Table 3.4. One can find that only a few cases

Table 3.3 A summary of system architectures.

System	Reference	System architecture of Smart PSS or similar systems
Smart PSS	Liu et al. (2018)	L1: Smart devices (embedded OS and physical components); L2: Network (transports of network equipment and transmission standards); L3: Data management (data storage and processing); L4: Applications (connect and interoperate the service platform with application programs)
Smart PSS	Zheng and Chen (2018)	L1: SCP layer (a number of smart devices connected in an IoT environment); L2: Edge computing layer (local data streams processing, and service matching/providing from the cloud to users with low latency); L3: Software-defined network management layer L4: Cloud computing layer (big data analytics with high latency)
An industrial IoT system architecture-based services framework	Maleki, Belkadi, Boli, et al. (2018)	L1: Edge tier (raw data collected from IoT devices, sensor data storage locally and edge implementation of services can be applied to this data) L2: Platform tier (services composition for operating in the cloud) L3: Enterprise tier (user access to cloud services' configuration, data, and outcomes)

perform a result-oriented business model (software-oriented), while most companies still adopt a product-oriented (e.g., remote diagnosis; Mourtzis, Vlachou, & Zogopoulos, 2017) or use-oriented one in a shared manner (e.g., cloud manufacturing; Charro & Schaefer, 2018). This is also in line with the fact that Smart PSS is enabled by the SCPs and its generated services in a servitization process rather than a productization manner.

30 Smart Product-Service Systems

Table 3.4 Eight types of PSS business models.

Business models	Types	Example	PSS description
Product-oriented	Product-related services: "the provider not only sells a product, but also offers services that are needed during the use phase of the product."	Smart wearable mask	The company sells the mask and provides the service of performance optimization according to the usage status.
	Advice and consultancy: "in relation to the product sold, the provider gives advice on its most efficient use."	Worry-Free Mobility for Electric Vehicles	Battery mobility advisor and battery life management analytics to improve the product for redesign purposes thereby bridging the battery back to business (Battery to Business).
Use-oriented	Product lease: "The provider has ownership, and is also often responsible for maintenance, repair and control. The lessee pays a regular fee for the use of the product."	Xerox Printer and Copier Rental	Providing world-class portfolio of copiers, printers, multifunction devices and production equipment to customers and a team of specialists to help with planning, installation, configuration and operation for the printer and copier rental.
		Rolls Royce "Power-by-the-Hour" service	A new service offering harnesses the power of Big Data to monitor, plan and perform maintenance and repairs on all the equipment installed on the cargo vessels.

Fundamentals of smart product-service system   31

Table 3.4 Eight types of PSS business models.—cont'd

Business models	Types	Example	PSS description
	Renting or sharing: "The same product is sequentially used by different users."	Car2Go	Offering car rental services but users can book the car on apps and leave the car anywhere on the street in the operating area.
	Product pooling: "a simultaneous use of the product."	3D Hubs	A cloud-based on-demand 3D printing service with a shared pool of manufacturing resources
Result-oriented	Activity management/outsourcing: "a part of an activity of a company is outsourced to a third party."	Bosch production line performance	Bosch crowdsources the Big Data analytics task of smart production line performance to the general public to seek best solution.
	Pay per service unit: "The PSS still has a fairly common product as a basis, but the user no longer buys the product, only the output of the product according to the level of use."	Air purification solutions	The users only care about the air purification service provided by the air purifier company with its own selection of products.
	Functional result: "for a functional result in rather abstract terms, which is not directly related to a specific technological system. The provider is, in principle, completely free as to how to deliver the result."	SF Express smart delivery service	SF Express has the "in 24-hour" transportation service for users caring about the service only.

Nevertheless, it is claimed that once the coverage/popularity of affordable SCPs reaches a certain level, they will be only considered as the common tools and media for service generation, and result-oriented business model (e.g., user experience rather than physical products) will be dominant by that time.

Furthermore, this book will provide more industrial examples of Smart PSS implementations of various business models in Chapter 7.

3.3.2 Digital platform

Smart PSS normally undertakes a platform-based approach (Cenamor, Rönnberg Sjödin, & Parida, 2017), which can be viewed as an organizational business perspective for leveraging the value of digital technologies based on modularity and IT-enabled interactions (Thomas, Autio, & Gann, 2014). Based on the types of services offered, the digital platforms of Smart PSS can be classified into two categories, i.e., *product-dependent digital platform* (digitalized services), and *product-independent digital platforms* (e-services), which are defined in Table 3.5, together with the typical studies in each category listed as well.

The *product-dependent digital platform* offers modularized digitalized services highly related to the physical product itself, such as product re-/configuration, real-time monitoring, and control. Meanwhile, the *product-independent digital platform* offers those modularized add-on e-services in the service platform which are independent with the functionality and status of the physical products, such as e-weather information, e-book, etc.

As mentioned before, this book will mainly address the prior type of digital platform following our narrowed definition of Smart PSS in the following chapters.

3.3.3 Value co-creation

Smart PSS follows the S–D logic (Vargo & Lusch, 2004), which represents a dynamic, continuing narrative of value co-creation through resource integration, and service exchange (Rönnberg Sjödin, Parida, & Wincent, 2016), enabled by the advanced IT. This value co-creation process (Liu et al., 2019; Mostafa, 2015) can be further depicted by its *interactive actors* (*coexist stakeholders*), *stages* (*co-design, co-implement, co-evaluate*), *mechanism* (*perceptive, responsive*), and *open innovation levels* (*product-service level, system level, system-of-systems level*), as shown in Table 3.6.

Table 3.5 Classification of digital platforms.

Categories	Definition	Types	Author, year	Examples
Product-dependent	An IoT-enabled modular platform consisting of various digitalized services.	Product re-/configuration, remanufacturing, and recycling system	Abramovici, Göbel, and Savarino (2017) Savarino, Abramovici, Göbel, and Gebus (2018) Zheng, Lin, Chen, and Xu (2018a), Zheng, Lin, Chen, & Xu (2018b) Hasselblatt, Huikkola, Kohtamäki, and Nickell (2018) Alcayaga, Wiener, and Hansen (2019) Zheng, Chen, and Shang (2019)	Configurable components of bicycle; reconfiguration of automobile components; remanufacturing of engine components, and etc.

Continued

Table 3.5 Classification of digital platforms.—cont'd

Categories	Definition	Types	Author, year	Examples
		Embedded open toolkits	Bénade, Brun, Le Masson, and Weil (2016) Zheng, Xu, and Chen (2018)	Self-adaptable shoes or wearables
		CPS/Digital Twin	Lee, Kao, and Yang (2014) Lee, Bagheri, and Kao (2015) Schroeder, Steinmetz, Pereira, and Espindola (2016) Schleich, Anwer, Mathieu, and Wartzack (2017), Filho, Liao, Loures, and Canciglieri (2017), Tao and Zhang (2017) Uhlemann, Lehmann, and Steinhilper (2017)	Monitoring, control, and optimization of machine tools; prognostic health management; high-fidelity design simulation, etc.

Product-independent	An IoT-enabled modular platform consisting of various e-services	Service platform	Söderberg, Wärmefjord, Carlson, and Lindkvist (2017) Tao, Zhang, Liu, and Nee (2018) Zheng et al. (2018)	
			Thomas et al. (2014) Lee and Kao (2014) Baines and Lightfoot (2014) Weiß, Kölmel, and Bulander (2016) Zhang, Liu, Liu, and Li (2016) Zheng et al. (2017) Cenamor, Sjödin, Parida (2017) Zheng et al. (2018a) Liu et al. (2018), Liu, Ming, and Song (2019) West, Gaiardelli, and Rapaccini (2018) Lee, Chen, and Trappey (2019)	Smart logistics arrangement; smart cooking advice; manufacturing capability planning; smart warning of breathing condition, etc.

Table 3.6 Key aspects of value co-creation in literature.

ID	References	Actors	Cocreation stages	Mechanisms	Open innovation level
1	Ana, Mugge, Schoormans, and Schifferstein (2014), Valencia et al. (2014, 2015)	Service provider and consumers	Co-design and evaluate	Perceptive, Responsive	System level
2	Takenaka et al. (2016)	Service provider and consumers	Co-evaluate	Responsive	System level
3	Weiß et al. (2016)	Suppliers and manufacturing companies	Co-implement	Perceptive	System-of–Systems level
4	Bénade et al. (2016)	User and designer	Co-design and implement	Perceptive	Product-Service level
5	Marilungo, Coscia, Quaglia, Peruzzini, and Germani (2016)	Platform provider (manufacturer) and end-users	Co-implement	Perceptive, Responsive	System-of–Systems level
6	Grubic and Peppard (2016)	Manufacturer, service provider and customer	Co-evaluation	Perceptive	System level
7	Lenka et al. (2017)	Manufacturer, service provider and customer	/	Perceptive, Responsive	/

8	Li and Found (2017)	Suppliers, customers, end-users and the society	/	/	System-of-Systems level
9	Liu et al. (2018, 2019)	Manufacturer, service provider and customer	Co-design, implement and evaluate	Perceptive, Responsive	System-of-Systems level
10	West et al. (2018)	Platform provider (manufacturer) and end-users/suppliers	Co-implement	Perceptive, Responsive	System-of-Systems level
11	Chowdhury, Haftor, and Pashkevich (2018)	Manufacturer, service provider and customer	Co-design, implement and evaluate	/	/
12	Zheng et al. (2018a, 2018b)	Manufacturer/designer, service provider, and users	Co-design, implement and evaluate	Perceptive, Responsive	Product-Service level and System level

One can find that in line with the broad scope of Smart PSS illustrated in Chapter 3.1, the level of open innovation corresponds to the interactive actors, where service provider/manufacturer are interacted in the first two levels to derive individualized solutions, while platform providers and manufacturing companies/suppliers interacted in the last level to optimize the overall system impact.

Meanwhile, among the cocreation stages, codesign stands for the active engagement of users in the value design process (e.g., smart configuration Bénade et al., 2016), coimplement represents the value implementation process through direct interaction and agent interaction (Liu et al., 2018) (e.g., platform provider and suppliers West et al., 2018), and coevaluate denotes the evaluation indexes including system-related (e.g., service performance), context-related (e.g., environmental information) and human-related (e.g., user experience) ones.

Lastly, enabled by the digital capabilities, the value co-creation mechanism can be classified into *perceptive mechanism* which allows the companies to identify, assess, and address specific customer needs (e.g., customize solutions in virtual space Zheng et al., 2018a), and *responsive mechanism,* which implies companies react to their customers' emerging and changing needs so that they can participate in the value co-creation quickly and proactively (e.g., dynamic pricing based on usage data logs; Takenaka, Yamamoto, Fukuda, Kimura, & Ueda, 2016).

In this book, the main issues of IT-driven value co-creation, including the role of user/customer, incentive mechanism, technical support and data-driven analytics will be further illustrated in Chapter 5.

3.4 Fundamentals of Smart PSS development

3.4.1 Two ways of development process

A systematic Smart PSS development process is of paramount importance to the final success of smart product-service provision, of which the main approaches are outlined in Table 3.7. One can find that the focus lies in two aspects: (1) *data-driven platform-based process*; (2) *value-driven cocreation process*, as depicted in Fig. 3.3.

For the prior one, it is conducted from a product-service provision generation perspective (Hagen, Kammler, & Thomas, 2018, pp. 87−99; Maleki, Belkadi, & Bernard, 2018), where massive user-generated and product-sensed data are collected mainly through SCPs and further analyzed in the service platform for Smart PSS provision generation. For

Table 3.7 Examples of Smart PSS development process.

Task	Ref.	Smart PSS development process
Smart PSS development process	Zheng et al. (2018a)	(1) Platform development; (2) Data acquisition and preprocessing; (3) Data analytics for service innovation; (4) Digital-twin-enabled service innovation.
Novel PSS development process	Scholze et al. (2016)	(1) Idea creation; (2) social simulator; (3) knowledge specification/data mining; (4) data capture/cyber-physical features selection; (5) functional specification determination; (6) context modeling; (7) security configuration; (8) product extension services orchestration; (9) PES development.
Smart service development	Verdugo Cedeño, Papinniemi, Hannola, and Donoghue (2018)	(1) Data from smart equipment; (2) customer needs assessment; (3) data analytics and decision support; (4) business model and value creation; (5) smart service selection.
Value-oriented Smart PSS development	Liu et al. (2018)	(1) Coexist the value flow of stakeholder. (2) Codesign the value proposition. (3) Coimplement the interactive value. (4) Coevaluate the performance value.
SPSE development process	Zheng, Gu, et al., (2017), Zheng, Ming, et al., (2017)	(1) Define (demand system boundary and customer demands); (2) Discover (discover typology model, robust mechanism, value emergency); (3) Design (design product clustering, service flow and service recommendation); (4) Delivery (capability planning, process management, and performance assessment).
Smart PSS design process	Lee et al. (2019)	(1) Problem definition: analyzing the service context and product characteristics, (2) resolution generation: generating service resolutions the Smart PSS solution, and (3) resolution design: smart PSS modeling.

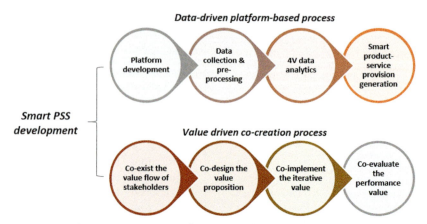

Figure 3.3 Two ways of Smart PSS development process.

example Zheng et al. (2018a) addressed a data-driven digital twin-enabled servitization process, and Scholze, Correia, and Stokic (2016) extended the scope with the consideration of knowledge management, context modeling, and security configuration, improving the quality and reliability of the generated product-service provision.

For the latter one, it is investigated from a value-driven perspective, where the interactions among stakeholders for value generation plays a dominant role. For instance, Liu et al. (2018) proposed a four-stage value co-creation process, of which both service providers and users are actively engaged in the so-called interactive activity diagrams. Meanwhile, Parida, Sjödin, and Reim (2019) articulated the components of value creation, value delivery, and value capture in the digital servitization process to reveal its significance in ensuring the success.

It is claimed that both approaches are considered in the Smart PSS development process in the following Chapters 4−9 consistently.

3.4.2 Three key characteristics

Based on the previous description, it can be found that unlike conventional design process starts from the very beginning of life-cycle, Smart PSS solution (product-service) design can be regarded as a value-generation process by considering both value creation at the early development stage and value recreation in the usage stage, as the *hybrid design* (Hou & Jiao, 2019). Meanwhile, SCPs own the powerful computation and communication capabilities with various built-in sensors that allow them to generate data and communicate to the Internet (Zhang, Member, et al., 2016). Hence, massive user-generated and product-sensed data can be leveraged, as the *hybrid intelligence*, to extract/generate valuable solution design concept. Moreover, as an IT-driven business strategy, both technical and business innovation, as the *hybrid value*, should be considered integrally (Zheng et al., 2018a). Hence, three unique characteristics of Smart PSS development can be outlined, including closed-loop design, IT-driven value co-creation, and design with context-awareness, in a data-driven manner (Cong et al., 2020). The interrelationship between design characteristics and key elements of Smart PSS is further depicted in Fig. 3.4.

Closed-loop design. Smart PSS takes both design and inverse design process into a close-loop consideration by utilizing the advanced IT, where informatic-based approaches serve as the key to enable its solution design and iteration success. Nevertheless, two challenges remain not well-solved at present. Firstly, in the early design phase, other than conventional PSS which provides a predefined product-service family in the web-based platform for user's configuration, Smart PSS should also well-define the smart, connectivity components in the IoT-enabled environment and be adaptable to the dynamic changes of user requirements during usage stage. Secondly, in the usage stage, driven by the massive user-generated data and

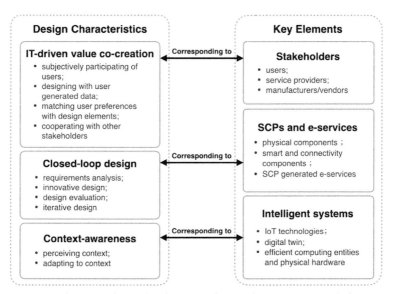

Figure 3.4 The interrelationship between design characteristics and elements.

product-sensed data in the smart, connected environment, Smart PSS should be capable of changing/upgrading with context-awareness for redesign innovation/improvement (Wang, Blache, Zheng, & Xu, 2018).

Therefore, a proper design methodology of Smart PSS should be explored to assist the stakeholders in completing not only the creation from scratch but also the real-time upgrade/modifications of product/service in the usage stage. Based on the four phases proposed by Liu et al. (2018), the four phases of Smart PSS closed-loop design are summarized in the following. (1) Requirements analysis phase. The user requirements, which deserve to be further addressed in Smart PSS, should be identified, collected, and analyzed in this phase (Hou & Jiao, 2019). (2) Innovative design phase. At this stage, the generation of new prototypes gets more attention. Some design methods can be used to output the innovative solutions which fulfill user requirements (Zheng et al., 2018a). (3) Design evaluation phase. The evaluation of Smart PSS could be conducted through three perspectives, including the customer value perspective (Qu et al., 2016), sustainability perspective (Liu et al., 2020), and value propositions perspective (Liu et al., 2019). (4) Iterative design phase. Smart PSS should quickly and automatically iterate its design plan to adapt to a new context when customers are using it. Some reasonable manners, such as changing/upgrading modules or controlling parameters, can become critical instruments in the iteration of SCPs and e-services to extend the lifespan of Smart PSS.

IT-driven value co-creation. Similar to the concept of open innovation 2.0 (Curley & Salmelin, 2013, pp. 1−12), Smart PSS is a value co-creation business paradigm by leveraging the cutting-edge digital technologies. Meanwhile, from industrial perspective, it can be seen as a sociotechnical ecosystem, where users become an integral part of the innovation process (i.e., cocreation) and their experience, i.e., user experience, drives the innovation.

The value co-creation process is carried out by stakeholders who are mainly classified into three species, i.e., users, service providers, and manufacturers/vendors (Zheng, Wang et al., 2019). Creating common values with IT is an expedient manner for Smart PSS design. Lenka et al. (2017) proposed a framework for digitalization capabilities that enabled value co-creation, which consists of the customer sphere, joint sphere, and provider sphere. The joint sphere is expanded by perceptive mechanisms and responsive mechanisms. Based on the above framework, the critical elements of IT-driven value co-creation are summarized in four aspects:

(1) For the customer sphere, customers should participate in the development process of Smart PSS subjectively through some appropriate approaches and technologies (Liu et al., 2018). Smart PSS should adopt a design methodology that can assist users in creating value independently by leveraging ICT.

(2) For the perceptive joint sphere, user-generated data should be utilized in Smart PSS design. Precepting users' preferences through collecting and processing real-time data are the most crucial part of the design process, for the reason that ensuring the real-time interaction between users and developers in the context is fundamental to the development of Smart PSS (Lenka et al., 2017).

(3) For the responsive joint sphere, user preferences should be matched with the design elements of Smart PSS in specific usage contexts. To respond to the collected users' preference, developers should construct some models to connect preferences (e.g., affective responses, desired impressions, and user types) to related design elements (e.g., product form features, patterns, and attributes) by leveraging artificial intelligence (Zheng et al., 2019b).

(4) For the service provider sphere, an IT-driven cooperation manner for the stakeholders should be concerned about in Smart PSS development (Li & Found, 2017). Scientific cooperation between these stakeholders in the context can effectively improve the development efficiency of Smart PSS.

Design with context-awareness is based on those intelligent systems (Wang et al., 2018), and defined as a wide range of technologies ensuring a high degree of connectivity and intelligence, such as IoT (Rymaszewska et al., 2017), DT (Schleich et al., 2017), computing entities, and physical hardware (Monostori et al., 2016). Meanwhile, product-service level design should follow adaptable design principles (Zhang, Gu, Peng, & Hu, 2017), but broaden its scope to both solution design considerations, by adding or replacing certain product-service modules through a predefined interface (e.g., API). Functional modeling is the key enabler for each design module to be independent of the others, so that the SCP and its e-services can be easily upgraded or changed based on individual user needs. Moreover, the solution design should be self-adaptable in the smart, connected environment with context-awareness (Bénade et al., 2016) to allow a certain degree of customer design freedom by embedding knowledge and rules about possible product or e-service differentiations into the SCP after manufactured, which is named *Smart PSS with built-in flexibility*. The end solution can be adjusted continuously during its life-cycle usage in a data-driven manner.

Wang et al. (2019) proposed that product-sensed data and user-generated data in the Smart PSS context should be primarily collected to enable real understanding of user behavior and trigger development. Based on the above statement, the focus of Smart PSS context-awareness lies in two perspectives: (1) perceiving context; and (2) adapting to context. For the prior one, intelligent systems help Smart PSS determining the present context with the hardware sensors which provide product-sensed information or the social sensors (Xu et al., 2018), which provide user-generated information on a social network. For the latter one, Smart PSS should update design solutions according to the contexts automatically or by the development team getting involved. Especially in the usage stage, the design approach driven by massive user-generated data in the smart, connected environment should provide tools for changing/upgrading the product/service predictively to adapt to the specific context (Zheng et al., 2019a).

3.4.3 Sustainability concerns

According to the definition of Smart PSS, the ultimate goal of Smart PSS is to meet individual customer needs with sustainability concerns. With the prevailing discussion and tendency toward circular economy (Tukker, 2015), it is also interesting to discover the interrelationship between digitalization capabilities and circular economy.

The authors extract the key finding in the digitalization capabilities for the circular economy from (Alcayaga et al., 2019; Bressanelli, Adrodegari, Perona, & Saccani, 2018; Opazo-Basáez, Vendrell-Herrero, & Bustinza, 2018; Pagoropoulos, Pigosso, & McAloone, 2017; Parida et al., 2019; Sinclair, Sheldrick, Moreno, & Dewberry, 2018) and further summarize in Table 3.8 with respect to *life-cycle stages, functionality, digitalization capabilities*, and *circular economy value driver*.

For a typical engineering life-cycle, it mainly consists of design, manufacturing, distribution/logistics, operation/use, and end-of-life stages. Meanwhile, the widely accepted circular economy value driver include *extended lifespan, increased resource efficiency*, and *close-the-loop*. One can find that from circular economy value driver perspective, it can be noted that most functionalities help extend product lifespan during design and usage stage (e.g., software upgrade) or achieve better resource efficiency during manufacturing stage (e.g., remote control), while only a few can close the loop during design or end-of-life stages (e.g., smart recycling, reuse, and remanufacturing).

Table 3.8 Digitalization capabilities for the circular economy.

Life-cycle stage	Functionality	Digitalization capabilities			Circular economy value driver		
		Connect capability	Intelligence capability	Analytic capability	Increased resource efficiency	Extended lifespan	Closed the loop
Design	Adaptable SCP design	✔	✔	✔		✔	✔
	Service innovation	✔	✔	✔		✔	
Manufacturing	Resource management	✔	✔	✔	✔		
	Production planning	✔	✔	✔	✔		
	Production process control/reconfiguration	✔	✔		✔		
	Geometry assurance	✔	✔	✔	✔		
Distribution/ logistic	Logistics and packaging	✔	✔	✔	✔		
Usage/ operation	Remote monitoring	✔	✔			✔	
	Predictive maintenance	✔	✔	✔		✔	
	Engineering change management	✔	✔	✔	✔	✔	
	Product reconfiguration	✔	✔	✔		✔	
	Product reuse	✔	✔			✔	
	Product remanufacturing	✔	✔	✔			✔
End-of-life	Product recycling	✔	✔	✔			✔

3.5 Chapter summary

This chapter introduced all the fundamentals of Smart PSS and its development to offer the readers an overview of this prevailing paradigm. A formal definition of Smart PSS, together with the declaration of misunderstanding concepts were introduced. Regarding to the technical aspect, Smart PSS is enabled by the disruptive digital technologies, with networking, intelligence, and analytic capabilities. Meanwhile, the four-layer based system architecture, including the physical resource layer, networked platform layer, service composition layer, and service application layer was depicted as the system foundation for Smart PSS development. For the business aspect, the eight types of Smart PSS business models were presented by following the original PSS principles. Meanwhile, both the product-dependent and product-independent digital platforms, and types of value co-creation mechanisms were summarized as well. Lastly, three unique characteristics, two fundamental ways of development, and sustainable concerns of Smart PSS in today's circular economy were illustrated. As the prerequisite for Smart PSS development, the main development challenges have been outlined, and will be further addressed in the following chapters accordingly.

References

Abramovici, M., Göbel, J. C., & Savarino, P. (2017). Reconfiguration of smart products during their use phase based on virtual product twins. *CIRP Annals - Manufacturing Technology, 66*(1), 165–168. https://doi.org/10.1016/j.cirp.2017.04.042.

Alcayaga, A., Hansen, E. G., & Wiener, M. (2019). Towards a framework of smart-circular systems: An integrative literature review. *Journal of Cleaner Production, 221*, 622–634. https://doi.org/10.1016/j.jclepro.2019.02.085.

Ana, V., Mugge, R., Schoormans, J. P. L., & Schifferstein, H. N. J. (September 2014). Challenges in the design of smart product-service systems (PSSs): Experiences from practitioners. In *19th DMI: Academic design management conference* (pp. 1–21).

Ardolino, M., Rapaccini, M., Saccani, N., Gaiardelli, P., Crespi, G., & Ruggeri, C. (2018). The role of digital technologies for the service transformation of industrial companies. *International Journal of Production Research, 56*(6), 2116–2132. https://doi.org/10.1080/00207543.2017.1324224.

Ardolino, M., Saccani, N., Gaiardelli, P., & Rapaccini, M. (2016). Exploring the key enabling role of digital technologies for PSS offerings. *Procedia CIRP, 47*, 561–566. https://doi.org/10.1016/j.procir.2016.03.238.

Baines, T., & Lightfoot, H. W. (2014). Servitization of the manufacturing firm: Exploring the operations practices and technologies that deliver advanced services. *International Journal of Operations and Production Management, 34*(1), 2–35. https://doi.org/10.1108/IJOPM-02-2012-0086.

Bénade, M., Brun, J., Le Masson, P., & Weil, B. (November 2016). How smart products with built in flexibility empower users to self - design the use: A theoretical framework for use generation. In *14th open and user innovation conference, Boston, United-States*. Retrieved from: https://hal.archives-ouvertes.fr/hal-01389650.

Brehm, L., & Klein, B. (2017). Applying the research on product-service systems to smart and connected products. In W. Abramowicz, R. Alt, & B. Franczyk (Eds.), *Business information systems workshops* (Vol. 263, pp. 311−319). Cham: Springer International Publishing.

Bressanelli, G., Adrodegari, F., Perona, M., & Saccani, N. (2018). Exploring how usage-focused business models enable circular economy through digital technologies. *Sustainability, 10*(3). https://doi.org/10.3390/su10030639.

Cenamor, J., Rönnberg Sjödin, D., & Parida, V. (2017). Adopting a platform approach in servitization: Leveraging the value of digitalization. *International Journal of Production Economics, 192*(December 2016), 54−65. https://doi.org/10.1016/j.ijpe.2016.12.033.

Charro, A., & Schaefer, D. (2018). Cloud manufacturing as a new type of product-service system. *International Journal of Computer Integrated Manufacturing, 31*(10), 1018−1033. https://doi.org/10.1080/0951192X.2018.1493228.

Chowdhury, S., Haftor, D., & Pashkevich, N. (2018). Smart product-service systems (smart PSS) in industrial firms: A literature review. *Procedia CIRP, 73*, 26−31. https://doi.org/10.1016/j.procir.2018.03.333.

Cong, J.-c., Chen, C.-H., Zheng, P., Li, X., & Wang, Z. (2020). A holistic relook at engineering design methodologies for smart product-service systems development. *Journal of Cleaner Production, 272*, 122737. https://doi.org/10.1016/j.jclepro.2020.122737.

Coreynen, W., Matthyssens, P., & Van Bockhaven, W. (2017). Boosting servitization through digitization: Pathways and dynamic resource configurations for manufacturers. *Industrial Marketing Management, 60*, 42−53. https://doi.org/10.1016/j.indmarman.2016.04.012.

Curley, M., & Salmelin, B. (2013). *Open innovation 2.0: A new paradigm*. OISPG White Paper. https://doi.org/10.1109/HIS.2008.172.

Filho, M. F., Liao, Y., Loures, E. R., & Canciglieri, O. (June 2017). Self-aware smart products: Systematic literature review, conceptual design and prototype implementation. *Procedia Manufacturing, 11*, 1471−1480. https://doi.org/10.1016/j.promfg.2017.07.278.

Grieves, M. (2014). *Digital twin: Manufacturing excellence through virtual factory replication*. Retrieved from: http://innovate.fit.edu/plm/documents/doc_mgr/912/1411.0_Digital_Twin_White_Paper_Dr_Grieves.pdf.

Grubic, T., & Peppard, J. (2016). Servitized manufacturing firms competing through remote monitoring technology an exploratory study. *Journal of Manufacturing Technology Management, 27*(2), 154−184. https://doi.org/10.1108/JMTM-05-2014-0061.

Hagen, S., Kammler, F., & Thomas, O. (2018). *Adapting product-service system methods for the digital era: Requirements for smart PSS engineering*. https://doi.org/10.1007/978-3-319-77556-2_6.

Hasselblatt, M., Huikkola, T., Kohtamäki, M., & Nickell, D. (2018). Modeling manufacturer's capabilities for the Internet of Things. *Journal of Business and Industrial Marketing, 33*(6), 822−836. https://doi.org/10.1108/JBIM-11-2015-0225.

Hou, L., & Jiao, R. J. (2019). Data-informed inverse design by product usage information: A review, framework and outlook. *Journal of Intelligent Manufacturing*. https://doi.org/10.1007/s10845-019-01463-2.

Kuhlenkötter, B., Bender, B., Wilkens, U., Abramovici, M., Göbel, J. C., Herzog, M., ... Lenkenhoff, K. (2017). Coping with the challenges of engineering smart product service systems - demands for research infrastructure. *Proceedings of the International Conference on Engineering Design, ICED, 3*(DS87−3), 341−350.

Kuhlenkötter, B., Wilkens, U., Bender, B., Abramovici, M., Süße, T., Göbel, J., ... Lenkenhoff, K. (2017). New perspectives for generating smart PSS solutions - life cycle, methodologies and transformation. *Procedia CIRP, 64*, 217–222. https://doi.org/10.1016/j.procir.2017.03.036.

Lee, E. A. (2008). Cyber physical systems: Design challenges. In *2008 11th IEEE international symposium on object and component-oriented real-time distributed computing (ISORC)*. https://doi.org/10.1109/ISORC.2008.25.

Lee, J., Bagheri, B., & Kao, H. A. (2015). A Cyber-Physical Systems architecture for Industry 4.0-based manufacturing systems. *Manufacturing Letters, 3*, 18–23. https://doi.org/10.1016/j.mfglet.2014.12.001.

Lee, C., Chen, C., & Trappey, A. J. C. (April 2019). Advanced Engineering Informatics A structural service innovation approach for designing smart product service systems: Case study of smart beauty service. *Advanced Engineering Informatics, 40*, 154–167. https://doi.org/10.1016/j.aei.2019.04.006.

Lee, J., & Kao, H. A. (2014). Dominant innovation design for smart products-service systems (PSS): Strategies and case studies. *Annual SRII Global Conference, SRII*, 305–310. https://doi.org/10.1109/SRII.2014.25.

Lee, J., Kao, H. A., & Yang, S. (2014). Service innovation and smart analytics for Industry 4.0 and big data environment. *Procedia CIRP, 16*, 3–8. https://doi.org/10.1016/j.procir.2014.02.001.

Lenka, S., Parida, V., & Wincent, J. (2017). Digitalization capabilities as enablers of value co-creation in servitizing firms. *Psychology and Marketing, 34*(1), 92–100. https://doi.org/10.1002/mar.

Lerch, C., & Gotsch, M. (2015). Digitalized product-service systems in manufacturing firms: A case study analysis. *Research-Technology Management, 58*(5), 45–52. https://doi.org/10.5437/08956308X5805357.

Li, A. Q., & Found, P. (2017). Towards sustainability: PSS, digital technology and value co-creation. *Procedia CIRP, 64*, 79–84. https://doi.org/10.1016/j.procir.2017.05.002.

Liu, Z., & Ming, X. (2019). A framework with revised rough-DEMATEL to capture and evaluate requirements for smart industrial product-service system of systems. *International Journal of Production Research, 0*(0), 1–19. https://doi.org/10.1080/00207543.2019.1577566.

Liu, Z., Ming, X., & Song, W. (2019). A framework integrating interval-valued hesitant fuzzy DEMATEL method to capture and evaluate co-creative value propositions for smart PSS. *Journal of Cleaner Production, 215*, 611–625. https://doi.org/10.1016/j.jclepro.2019.01.089.

Liu, Z., Ming, X., Song, W., Qiu, S., & Qu, Y. (2018). A perspective on value co-creation-oriented framework for smart product-service system. *Procedia CIRP, 73*, 155–160. https://doi.org/10.1016/j.procir.2018.04.021.

Liu, Z., Ming, X., Qiu, S., Qu, Y., & Zhang, X. (2020). A framework with hybrid approach to analyse system requirements of smart PSS toward customer needs and co-creative value propositions. *Computers & Industrial Engineering, 139*, 105776. https://doi.org/10.1016/j.cie.2019.03.040.

Maglio, P. P., & Lim, C.-H. (2016). Innovation and big data in smart service systems. *Journal of Innovation Management, 4*(1), 11–21.

Maleki, E., Belkadi, F., & Bernard, A. (2018). Industrial product-service system modelling base on systems engineering: Application of sensor integration to support smart services. *IFAC-PapersOnLine, 51*(11), 1586–1591. https://doi.org/10.1016/j.ifacol.2018.08.270.

Maleki, E., Belkadi, F., Boli, N., van der Zwaag, B. J., Alexopoulos, K., Koukas, S., ... Mourtzis, D. (2018). Ontology-based framework enabling smart product-service systems: Application of sensing systems for machine health monitoring. *IEEE Internet of Things Journal, 5*(6), 4496–4505. https://doi.org/10.1109/JIOT.2018.2831279.

Marilungo, E., Coscia, E., Quaglia, A., Peruzzini, M., & Germani, M. (2016). Open innovation for ideating and designing new product service systems. *Procedia CIRP, 47,* 305–310. https://doi.org/10.1016/j.procir.2016.03.214.

Monostori, L., Kádár, B., Bauernhansl, T., Kondoh, S., Kumara, S., Reinhart, G., ... Ueda, K. (2016). Cyber-physical systems in manufacturing. *CIRP Annals, 65*(2), 621–641. https://doi.org/10.1016/j.cirp.2016.06.005.

Mostafa, R. B. (2015). Value co-creation in industrial cities: A strategic source of competitive advantages. *Journal of Strategic Marketing, 24*(2), 144–167. https://doi.org/10.1080/0965254X.2015.1076885.

Mourtzis, D., Vlachou, A., & Zogopoulos, V. (2017). Cloud-based augmented reality remote maintenance through shop-floor monitoring: A product-service system Approach. *Journal of Manufacturing Science and Engineering, 139*(6), 061011. https://doi.org/10.1115/1.4035721.

Mourtzis, D., Zogopoulos, V., & Vlachou, E. (2017). Augmented reality application to support remote maintenance as a service in the robotics industry. *Procedia CIRP, 63,* 46–51. https://doi.org/10.1016/j.procir.2017.03.154.

Opazo-Basáez, M., Vendrell-Herrero, F., & Bustinza, O. F. (2018). Uncovering productivity gains of digital and green servitization: Implications from the automotive industry. *Sustainability, 10*(5). https://doi.org/10.3390/su10051524.

Pagoropoulos, A., Pigosso, D. C. A., & McAloone, T. C. (2017). The emergent role of digital technologies in the circular economy: A review. *Procedia CIRP, 64,* 19–24. https://doi.org/10.1016/j.procir.2017.02.047.

Parida, V., Sjödin, D. R., Lenka, S., & Wincent, J. (2015). Developing global service innovation capabilities: How global manufacturers address the challenges of market heterogeneity. *Research-Technology Management, 58*(5), 35–44. https://doi.org/10.5437/08956308x5805360.

Parida, V., Sjödin, D., & Reim, W. (2019). Reviewing literature on digitalization, business model innovation, and sustainable industry: Past achievements and future promises. *Sustainability, 11*(2). https://doi.org/10.3390/su11020391.

Porter, M. E., & Heppelmann, J. E. (2014). How smart, connected products are transforming competition. *Harvard Business Review, 92*(11), 64–88.

Qu, M., Yu, S., Chen, D., Chu, J., & Tian, B. (2016). State-of-the-art of design, evaluation, and operation methodologies in product service systems. *Computers in Industry, 77,* 1–14. https://doi.org/10.1016/j.compind.2015.12.004.

Rönnberg Sjödin, D., Parida, V., & Wincent, J. (2016). Value co-creation process of integrated product-services: Effect of role ambiguities and relational coping strategies. *Industrial Marketing Management, 56,* 108–119. https://doi.org/10.1016/j.indmarman.2016.03.013.

Rymaszewska, A., Helo, P., & Gunasekaran, A. (2017). IoT powered servitization of manufacturing – an exploratory case study. *International Journal of Production Economics, 192,* 92–105. https://doi.org/10.1016/j.ijpe.2017.02.016.

Savarino, P., Abramovici, M., Göbel, J. C., & Gebus, P. (2018). Design for reconfiguration as fundamental aspect of smart products. *Procedia CIRP, 70,* 374–379. https://doi.org/10.1016/J.PROCIR.2018.01.007.

Schleich, B., Anwer, N., Mathieu, L., & Wartzack, S. (2017). Shaping the digital twin for design and production engineering. *CIRP Annals - Manufacturing Technology, 66*(1), 141–144. https://doi.org/10.1016/j.cirp.2017.04.040.

Scholze, S., Correia, A. T., & Stokic, D. (2016). Novel tools for product-service system engineering. *Procedia CIRP, 47,* 120–125. https://doi.org/10.1016/j.procir.2016.03.237.

Schroeder, G. N., Steinmetz, C., Pereira, C. E., & Espindola, D. B. (2016). Digital twin data modeling with AutomationML and a communication methodology for data exchange. *IFAC-PapersOnLine, 49*(30), 12–17. https://doi.org/10.1016/j.ifacol.2016.11.115.

Sinclair, M., Sheldrick, L., Moreno, M., & Dewberry, E. (2018). Consumer intervention mapping-A tool for designing future product strategies within circular product service systems. *Sustainability, 10*(6). https://doi.org/10.3390/su10062088.

Söderberg, R., Wärmefjord, K., Carlson, J. S., & Lindkvist, L. (2017). Toward a digital twin for real-time geometry assurance in individualized production. *CIRP Annals - Manufacturing Technology, 66*(1), 137—140. https://doi.org/10.1016/j.cirp.2017.04.038.

Takenaka, T., Yamamoto, Y., Fukuda, K., Kimura, A., & Ueda, K. (2016). Enhancing products and services using smart appliance networks. *CIRP Annals - Manufacturing Technology, 65*(1), 397—400. https://doi.org/10.1016/j.cirp.2016.04.062.

Tao, F., Cheng, J., Qi, Q., Zhang, M., Zhang, H., & Sui, F. (2017). Digital twin-driven product design, manufacturing and service with big data. *International Journal of Advanced Manufacturing Technology*, 3563—3576. https://doi.org/10.1007/s00170-017-0233-1.

Tao, F., & Zhang, M. (2017). Digital twin shop-floor: A new shop-floor paradigm towards smart manufacturing. *IEEE Access, 5*, 20418—20427. https://doi.org/10.1109/ACCESS.2017.2756069.

Tao, F., Zhang, M., Liu, Y., & Nee, A. Y. C. (2018). Digital twin driven prognostics and health management for complex equipment. *CIRP Annals, 67*(1), 169—172. https://doi.org/10.1016/j.cirp.2018.04.055.

Thomas, L. D. W., Autio, E., & Gann, D. M. (2014). Architectural leverage: Putting platforms in context. *Academy of Management Perspectives, 28*(2), 198—219.

Tukker, A. (2004). Eight types of product-service system: Eight ways to sustainability? Experiences from suspronet. *Business Strategy and the Environment, 13*(4), 246—260. https://doi.org/10.1002/bse.414.

Tukker, A. (2015). Product services for a resource-efficient and circular economy — a review. *Journal of Cleaner Production, 97*, 76—91. https://doi.org/10.1016/j.jclepro.2013.11.049.

Uhlemann, T. H. J., Lehmann, C., & Steinhilper, R. (2017). The digital twin: Realizing the cyber-physical production system for industry 4.0. *Procedia CIRP, 61*, 335—340. https://doi.org/10.1016/j.procir.2016.11.152.

Valencia Cardona, A. M., Mugge, R., Schoormans, J. P. L., & Schifferstein, H. N. J. (2014). Challenges in the design of smart product-service systems (PSSs): Experiences from practitioners. In *Proceedings of the 19th DMI: Academic design management conference. Design management in an era of disruption, London, UK, September 2—4, 2014. Design Management Institute*.

Valencia, A., Mugge, R., Schoormans, J. P. L., & Schifferstein, H. N. J. (2015). The design of smart product-service systems (PSSs): An exploration of design characteristics. *International Journal of Design, 9*(1), 13—28. https://doi.org/10.1016/j.procir.2016.04.078.

Vargo, S. L., & Lusch, R. F. (2004). Evolving to a new dominant logic for marketing. *Journal of Marketing, 68*(1), 1—17. https://doi.org/10.1509/jmkg.68.1.1.24036.

Vendrell-Herrero, F., Bustinza, O. F., Parry, G., & Georgantzis, N. (2017). Servitization, digitization and supply chain interdependency. *Industrial Marketing Management, 60*, 69—81. https://doi.org/10.1016/j.indmarman.2016.06.013.

Verdugo Cedeño, J. M., Papinniemi, J., Hannola, L., & Donoghue, I. D. M. (2018). Developing smart services by Internet of things in manufacturing business. *DEStech Transactions on Engineering and Technology Research, 14*(icpr), 59—71. https://doi.org/10.12783/dtetr/icpr2017/17680.

Wang, Y., Blache, R., Zheng, P., & Xu, X. (2018). A knowledge management system to support design for additive manufacturing using Bayesian networks. *Journal of Mechanical Design, Transactions of the ASME, 140*(5). https://doi.org/10.1115/1.4039201.

Wang, Z., Chen, C.-H., Zheng, P., Li, X., & Khoo, L. P. (2019). A novel data-driven graph-based requirement elicitation framework in the smart product-service system context. *Advanced Engineering Informatics, 42,* 100983. https://doi.org/10.1016/j.aei.2019.100983.

Weiß, P., Kölmel, B., & Bulander, R. (2016). Digital service innovation and smart technologies: Developing digital strategies based on industry 4.0 and product service systems for the renewal energy sector. In *Proceedings of the 26th annual RESER conference, Naples, Italy* (pp. 274−291).

West, S., Gaiardelli, P., & Rapaccini, M. (2018). Exploring technology-driven service innovation in manufacturing firms through the lens of Service Dominant logic. *IFAC-PapersOnLine, 51*(11), 1317−1322. https://doi.org/10.1016/j.ifacol.2018.08.350.

Wiesner, S., & Thoben, K.-D. (2017). Cyber-physical product-service systems. In S. Biffl, A. Lüder, & D. Gerhard (Eds.), *Multi-disciplinary engineering for cyber-physical production systems* (pp. 63−88). Cham: Springer International Publishing.

Xu, Z., Mei, L., Choo, K.-K. R., Lv, Z., Hu, C., Luo, X., & Liu, Y. (2018). Mobile crowd sensing of human-like intelligence using social sensors: a survey. *Neurocomputing, 279,* 3−10.

Zhang, J., Gu, P., Peng, Q., & Hu, S. J. (2017). Open interface design for product personalization. *CIRP Annals - Manufacturing Technology, 66*(1), 173−176. https://doi.org/10.1016/j.cirp.2017.04.036.

Zhang, Y., Liu, S., Liu, Y., & Li, R. (2016). Smart box-enabled product−service system for cloud logistics. *International Journal of Production Research, 54*(22), 6693−6706. https://doi.org/10.1080/00207543.2015.1134840.

Zhang, X., Member, S., Yang, Z., Sun, W., Member, S., Liu, Y., & Tang, S. (2016). Incentives for mobile crowd sensing: A survey. *IEEE Communications Surveys and Tutorials, 18*(1), 54−67.

Zheng, P., & Chen, C.-H. (2018). A hybrid crowdsensing approach with cloud-edge computing framework for design innovation in smart product-service systems. In *48th international conference on computers and industrial engineering, Auckland, New Zealand.*

Zheng, P., Lin, T.-J., Chen, C.-H., & Xu, X. (2018a). A systematic design approach for service innovation of smart product-service systems. *Journal of Cleaner Production, 201,* 657−667. https://doi.org/10.1016/j.jclepro.2018.08.101.

Zheng, P., Lin, Y., Chen, C.-H., & Xu, X. (2018b). Smart, connected open architecture product: An IT-driven co-creation paradigm with lifecycle personalization concerns. *International Journal of Production Research, 0*(0), 1−14. https://doi.org/10.1080/00207543.2018.1530475.

Zheng, M., Ming, X., Wang, L., Yin, D., & Zhang, X. (2017). Status review and future perspectives on the framework of smart product service ecosystem. *Procedia CIRP, 64,* 181−186. https://doi.org/10.1016/j.procir.2017.03.037.

Zheng, P., Chen, C.-H., & Shang, S. (2019a). Towards an automatic engineering change management in smart product-service systems—A DSM-based learning approach. *Advanced Engineering Informatics, 39,* 203−213.

Zheng, P., Wang, Z., & Chen, C. H. (2019b). Industrial smart product-service systems solution design via hybrid concerns. *Procedia Cirp, 83,* 187−192.

Zheng, P., Xu, X., & Chen, C. H. (2020). A data-driven cyber-physical approach for personalised smart, connected product co-development in a cloud-based environment. *Journal of Intelligent Manufacturing, 31*(1), 3−18.

CHAPTER 4

Design entropy theory

Contents

4.1 Challenges of typical design methodologies	53
4.2 Design entropy theory	54
4.2.1 Fundamentals	59
4.2.2 Design entropy	62
4.2.3 Innovative design entropy	64
4.2.4 Iterative design entropy	66
4.3 Self-adaptable design process of smart PSS	70
4.4 Information conversion map tool	72
4.5 Case study	75
4.6 Summary	81
References	81

As outlined in Chapter 3, closed-loop, self-adaptable design; IT-driven value co-creation; and design with context-awareness are deemed as the three unique design characteristics of Smart PSS, conducted in a data-driven manner. In this context, unlike the conventional design process that starts from the very beginning of the lifecycle, Smart PSS design innovation can be regarded as a value generation process by considering the whole product-service lifecycle in a closed-loop manner with context-awareness. Hence, it is of paramount importance to provide a fundamental approach for Smart PSS development by considering its unique design characteristics holistically in the digital servitization era.

4.1 Challenges of typical design methodologies

In order to find any proper methodology from the existing databases, the concept of "smart design" (Zheng, Wang, Chen, & Khoo, 2019) is adopted to distinguish design methods for Smart PSS development from conventional ones, which is defined as the ones that: (1) the research objects are smart products, smart services, or Smart PSS; (2) leverage smart or digital technologies (e.g., artificial intelligence and IoT); or (3) adopted in smart areas (e.g., smart home) (Cong, Chen, & Zheng, 2020). Motivated by this, several engineering design methodologies that have been adopted in PSS

Smart Product-Service Systems
ISBN 978-0-323-85247-0
https://doi.org/10.1016/B978-0-323-85247-0.00002-5

© 2021 Elsevier Inc.
All rights reserved.

53

development and have been researched on "smart design" before are further identified based on the Scopus database, owing to its broad coverage of most important research works. Meanwhile, search terms such as "smart" and "digital" were also used in the search strategy to make search results more relevant, and the timespan starts from 1999 to 2020.

Through systematic review of the related articles, eight typical engineering design methodologies (i.e., TRIZ, quality function deployment (QFD), Kansei Engineering (KE), user-centered design (UCD), AD, Blueprint design, Adaptable design, and functional-behavior-structure (FBS)) are included. As shown in Table 4.1, most existing works have investigated Smart PSS development from either product or service aspect, respectively (Zheng et al., 2019). From the product-oriented aspect, the existing body of research suggests that some features (e.g., self-awareness (Filho, Liao, Loures, & Canciglieri, 2017) and reconfigurability (Savarino, Abramovici, Göbel, & Gebus, 2018)) frequently prescribe for designing SCPs. From the service-oriented aspect, several studies explore approaches based on IoT (Liu, Ming, & Song, 2019) or data-driven techniques for developing smart e-services (Verdugo Cedeño, Papinniemi, Hannola, & Donoghue, 2018), and many methods for advanced digitalized services (e.g., digital twin (Tao, Cheng, Qi, Zhang, Zhang, & Sui, 2017) and augmented reality (AR) (Gupta et al., 2018). Although the design method for Smart PSS development has received much attention recently, scarcely any work provides a fundamental approach to realizing Smart PSS development by considering its unique design characteristics (i.e., IT-driven value co-creation, closed-loop design, and context-awareness) in the digital servitization era (Liu, Ming, Qiu, Qu, & Zhang, 2020).

Meanwhile, the eight design methodologies are further categorized by its capability to meet any of the three unique key characteristics of Smart PSS defined in Chapter 3, as detailed in Table 4.2. It can be found that none of these methods can address these characteristics holistically (Cong, Chen, & Zheng, 2020). Aiming to fill this gap, this chapter presents a novel design entropy theory for the Smart PSS development, which takes the process of information prediction, summarization, conversion, and updating into an overall consideration.

4.2 Design entropy theory

In order to facilitate the Smart PSS development, this section proposes a new design methodology, i.e., design entropy theory (DET), by leveraging

Table 4.1 Definition and research objects of the eight design methodologies.

Design methodologies	Research objects		Definitions	
	Smart product	Smart service	Specification	Ref.
TRIZ	(Koswatte, Paik, Park, & Kumara, 2015; Moehrle, 2010; Wang, 2015; Wang, 2017)	(Lee, Wang, & Trappey, 2015; Lee, Chen, & Trappey, 2019; Wang and Chin, 2017)	TRIZ (from the Russian phrase "Teorija Rezhenija Izobretatelskih Zadach") proposed by Russian researcher is a creative problem-solving theory, which was developed as a knowledge-based innovative approach for solving conflicts in technical systems by some techniques.	Ilevbare, Probert and Phaal (2013), Savransky (2000)
QFD	(Li, Wang, & Wu, 2014; Choi, Kim, Lee, & Kwon, 2015; Wang et al., 2015; Wang, 2017; Kim, Sul, & Choi, 2018)	(Sohn, Kim, & Lee, 2013)	Quality function deployment (QFD) as a product development methodology driven by customer requirements is to decompose the implementation process of customer demands into different stages of product development, and evaluate the product performance according to the customer satisfaction.	Zairi and Youssef (1995), Köksal and Egitman (1998), Govers (2001)
KE	(Wang and Chin, 2017; Li, Tian, Wang, Wang, & Huang, 2018)	/	Kansei Engineering (KE) was developed initially in Japan as a consumer-oriented product design technology. Through studying the	Jindo and Hirasago (1997), Nagamachi (1995),

Table 4.1 Definition and research objects of the eight design methodologies.—cont'd

Design methodologies	Research objects		Definitions	
	Smart product	Smart service	Specification	Ref.
			emotional response of users to the product systematically, and transforming the emotion of customers into measurable physical design parameters by ergonomics and computer science.	Nagamachi and Imada (1995)
UCD	/	(Augusto, Kramer, Alegre, Covaci, & Santokhee, 2017)	The term user-centered design (UCD) originally came from Donald Norman's research laboratory, emphasizing the core role of user information during each phase of the design process and proposes that early user engagement can facilitate the design process effectively.	Anderson, Norman and Draper (1988), Karat and Watson (1996), Nugroho (2012)
AD	(Rauch, Dallasega, & Matt, 2016; Riel, Kreiner, Messnarz, & Much, 2018)	/	Axiomatic design (AD) was first proposed by Suh. AD theory submits two essential axioms, include the independence axiom which maintains the uncoupled of functional requirements, and the information axiom, which minimizes product information content that can eliminate the possibility of	Suh (1998, 2007)

			making mistakes in the product development process.	
Blueprint design	/	(Lee et al., 2015; Lee et al., 2019; Xu, 2020)	Service Blueprint proposed by Shostack (1982) is an analysis approach by integrating service content and information structure on a clear map to help designers investigate organizational service processes based on customer behavior.	Chou, Chen and Conley (2012), Shostack (1982)
Adaptable design	(Peng, Liu, Gu, & Fan, 2013; Zhang, Xue, & Gu, 2015; Hu, Peng, & Gu, 2015; Zheng, Xu, Yu, & Liu, 2017; Zheng, Xu, & Chen, 2018; Zheng, Lin, Chen and Xu 2019)	/	Adaptable design was first presented by Gu to adapt new requirements through design adaptability or product adaptability with replacing or adding certain modules through predefined adaptive interfaces when the conditions change.	Gu, Hashemian and Nee (2004), Gu, Xue and Nee (2009)
FBS	(Liu et al., 2019; Qin et al., 2019)	/	Functional-behavior-structure (FBS) framework was proposed by Gero (1990) to represent the design process as the translation between function, behavior, and structure. The design flow of the FBS framework was presented as eight steps.	Gero (1990), Gero and Kannengiesser (2004)

Derived from Cong, J. C. Chen, C. -H., Zheng, P., Li, X., & Wang, Z. (2020). A holistic relook at engineering design methodologies for smart product-service systems development. *Journal of Cleaner Production, 272*(June), 122737.

Table 4.2 Comparison of typical design methodologies for Smart PSS.

	IT-driven value co-creation				Closed-loop design				Context-awareness	
	Subjectively participating of users	Designing with user-generated data	Matching user preferences with design elements	Cooperating with other stakeholders	Requirements analysis	Innovative design	Design evaluation	Iterative design	Perceiving context	Adapting to context
TRIZ		✓	✓		✓	✓	✓			
QFD	✓	✓		✓	✓	✓	✓			
KE	✓	✓	✓				✓			
UCD	✓	✓		✓		✓	✓	✓	✓	
AD						✓				
Service blueprint						✓	✓			
Adaptable design	✓	✓	✓				✓	✓	✓	
FBS					✓	✓	✓		✓	

Derived from Cong, JC, Chen, C. H., & Zheng, P. (2020). Design entropy theory: A new design methodology for smart PSS development. *Advanced Engineering Informatics*.

Design entropy theory 59

the information theory to take the process of information prediction, summarization, conversion and updating into an overall consideration. Its fundamental issues, definition, design process, tool, and benefits are elaborated in the following sections.

4.2.1 Fundamentals

Before introducing DET, we must first understand what information is. Information is an existence independent of material and energy. To fully comprehend information, let us first clarify the concepts of data, information, and knowledge (Targowski, 2005):

Data: shows the observable attributes of objects, events, and their environment.

Information: consists of processed data, and organizes data into valuable content.

Knowledge: represents the abstraction, generalization, and application of information.

It should be noted, like data, information also depicts the attributes, but it does so more compactly and usefully than data. The difference between data and information is functional, not structural (Targowski, 2005). For example:

Heart rate: 100.

Can you see anything from this data? This should be not easy. Then we look at the following information:

Name: Tom.

Measure time: After exercise.

Heart rate: 100.

Hereby, 100 is meaningful. 100 represents Tom's heart rate after exercise and becomes a key indicator in meaningful information. Knowledge is the abstraction of information obtained by filtering, refining, and processing relevant information. For example:

The heart rate of 100 after exercise is healthy.

The above sentence is a piece of knowledge. In addition, new knowledge can be generated through reasoning and analysis. For example:

Knowledge 1: The user number of smart bicycles at 8 o'clock is higher than 9 o'clock.

Knowledge 2: The user number of smart bicycles at 9 o'clock is higher than 10 o'clock.

New knowledge: The user number of smart bicycles at 8 o'clock is higher than 10 o'clock.

In this case, new knowledge is derived from knowledge 1 and knowledge 2. Fig. 4.1 depicts the relationship hierarchy of data, information and

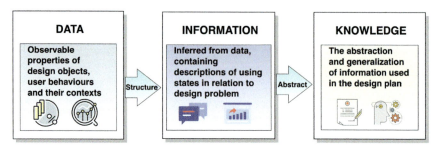

Figure 4.1 The relationship between three levels. *(Derived from Wodehouse, A. J. & W J. Ion. (2010). Information use in conceptual design: Existing taxonomies and new approaches.* International Journal of Design, *4(3), 53–65.)*

knowledge in the Smart PSS development domain based on the DIKW model, proposed by Wodehouse and Ion (2010). Information sits in between the other two levels, which can be applied in Smart PSS design process. The scope of information in Smart PSS needs to be clarified. Information in Smart PSS can be the descriptive and prescriptive one revealing the status of the stakeholders, solution (i.e., SCPs and services), and environment. The information can be input by stakeholders directly or extracted from hardware/social sensors collected data.

After understanding the concept of information, the basic idea of DET can be defined as (Cong, Chen, & Zheng, 2020):

DET does not regard designing physical products as a goal. DET considers that the information in the system is the design object, and information is stored in the container (i.e., Smart PSS and stakeholders). The design process of Smart PSS can be regarded as the process of collecting information from specific container or environment and transferring/ materializing it to other physical containers.

For example, from a traditional perspective, when we design a bicycle, we should focus on the size, shape, material, and color of this bicycle. However, from a DET perspective, when designing a smart bicycle service system, we should focus on its information activities (i.e., information collection, prediction, conversion, transmission, and updating). The smart bicycle service system can get the user physiological data and environmental data, such as the user's height data. Furthermore, the information extracted from data is that 80% of the users are taller than 1.8 m. Then this information will materialize to the corresponding solution on the physical level, e.g., the position of bicycle seat is adjusted higher.

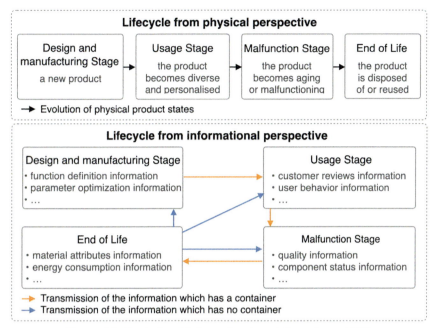

Figure 4.2 Physical and informational evolution of smart PSS.

Why DET does not choose physical substance but information as the design objective? Because the physical substance of Smart PSS will eventually be eliminated. Nevertheless, the information is unlike in nature. Information can be stored completely and transmitted to a new container when its old containers are destroyed. The lifecycles of Smart PSS from physical and informational perspectives are depicted in Fig. 4.2.

Let us take the smart electric bicycle service system as an example. From the physical perspective, a new smart electric bicycle is output after the design and manufacturing stage. This bicycle becomes more personalized during using, i.e., some users may install a child seat, and some users may install a transmission for the bicycle. After using for a long time, the hardware of the bicycle becomes worn out, resulting in aging and damage to the parts. At last, when the smart electric bicycle reaches the end of the lifecycle, the physical substance will be disposed.

However, if we observe the lifecycle from the informational perspective, information is continuously accumulated throughout the whole lifecycle. In the design and manufacturing stage, the system can collect the size, material, shape, color information of the bicycle, assembly information of parts, etc. In the usage stage, the system can get user evaluation and feedback information to the system, user behavior information (e.g., riding speed and riding route information), etc. In the usage stage, the system can collect fault cause information, maintenance service information, etc. At the end of life, the system can continue to collect some information (e.g., materials recycled way information). In addition, the information carried by the eliminated Smart PSS can be saved or transmitted to other containers.

4.2.2 Design entropy

DET, which takes information as the research object, is based on the concept of information entropy in information theory. Information theory was proposed by Shannon in 1949 (Shannon, 1949), which systematically discussed the fundamental issues of information communication by using mathematical tools (e.g., probability theory and statistics), including both the quantitative expression of information and the concept of information entropy. In information theory, information entropy plays a significant role to measure system uncertainty. The information entropy of a system H(x) is given by:

$$H(x) = -\sum_{i=1}^{n} p(x_i)\log(p(x_i)) \tag{4.1}$$

To understand this formula, we must first learn about the amount of information: it is a measure of information which is related to the occurrence probabilities of events. If the occurrence probability of the event is low, the amount of information is high, such as the information "Mary gave birth to quadruplets." (the probability of this happening is relatively low, so the amount of information is high). Meanwhile, if the occurrence probability of the event is high, the amount of information is low, such as the information "today has 24 h" (it always happens, so the amount of information is very low). Furthermore, how to express the amount of information with probability? There are two unrelated events, x and y. The amount of information when x and y occur at the same time is the sum of the amount of information when x appears alone and the amount of information when y appears alone:

$$h(x, \ y) = h(x) + h(y) \tag{4.2}$$

Since x and y are unrelated events, their probabilities of occurrence satisfy:

$$p(x,\ y) = p(x)*p(y) \tag{4.3}$$

In other words, $h(x)$ is related to the logarithm of $p(x)$. For the reason that multiply of the true number can be corresponded to the addition of the logarithm in the logarithm formula. The amount of information is given by:

$$h(x) = -\log(p(x)) \tag{4.4}$$

The amount of information measures the information brought after the specific event occurs. Furthermore, information entropy is the expectation of the amount of information from all possible events before the specific event occurs (See Formula 4.1). Information entropy can be used as a measure of system complexity. If the system is complex and has many different kinds of possible occurrence, its information entropy is relatively high. If a system is simple and has a little kind of possible occurrence (e.g., there is only one possible situation, then the probability is 1, and the information entropy is 0), the information entropy is relatively low.

Based on information entropy, design entropy can be "a measure of a deterministic degree of Smart PSS information design." The value of design entropy is positively related to the systems' uncertainty. If a system's certainty is low, its design entropy should be high, which means that the accuracy of its context-awareness, the quality of real-time iterative designing based on the current context, and the stakeholder's participation and satisfaction degree are low. Consequently, DET needs to dynamically reduce the design entropy in order to keep the most outstanding deterministic for each context throughout the lifecycle.

To calculate the design entropy, a new measurement called conversion ability is introduced, which is denoted by C. Conversion ability represents the ability to convert the information denoted by x_i into the information denoted by x'. As shown in Fig. 4.4, x refers to the information which has no solutions mapping, and x' refers to the information which describes design solutions. The conversion ability of Smart PSS is high, which means that its conversion effect from x to x' is excellent, and the design entropy of this system is low. Let D as the design entropy, and the value of D is calculated by:

$$D = -\log C \tag{4.5}$$

where $0 \leq C \leq 1$, and all $D \geq 0$. Therefore, if the conversion ability of a system is the highest, the value of C is 1, then $D = 0$. On the contrary, the

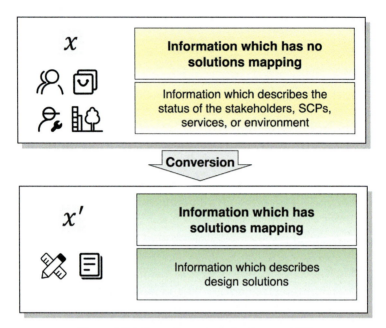

Figure 4.3 Information conversion in smart PSS.

value of C is 0, then $D = +\infty$. In order to compare the value of design entropy in different contexts and systems, the same logarithmic base 2 is used in each equation of design entropy. Furthermore, the total design entropy of a system is given by

$$D_{total} = D_{innovative} + D_{iterative} \qquad (4.6)$$

where $D_{innovative}$ represents innovative design entropy of the system, and $D_{iterative}$ represents iterative design entropy. The former one means the design entropy in the innovative design stage of creating a new Smart PSS. Furthermore, the latter one means the design entropy in the usage and iterative design stage at which the design solutions can be redesigned, modified, or enhanced according to the new information collected from sensors (Fig. 4.3).

4.2.3 Innovative design entropy

We define the innovative design entropy of a system as $D_{innovative}$ according to the innovative conversion of the system, $C_{innovative}$. Hence, $D_{innovative}$ is defined as:

$$D_{innovative} = -\log C_{innovative} \qquad (4.7)$$

This quantity measures how uncertain the system is in its innovative design stage when we know $C_{innovative}$. $C_{innovative}$ refers to the conversion ability of the design entropy to effectively convert information. That is

$$C_{innovative} = A_1 \times Y_1 + A_2 \times Y_2 + \cdots + A_n \times Y_n \qquad (4.8)$$

where Y_n is a parameter that represents the ability that a design entropy can convert the information in a specific way when the plan is adopted in the future. Since $0 \leq C_{innovative} \leq 1$, let $0 \leq Y_n \leq 1$. Y_n is the result of Z_n mapping to $[0, 1]$. To meet the three features of Smart PSS, the matching degree with user requirements is utilized as Z_1 to calculate Y_1 of Smart PSS. The adaptability measure is appropriated as Z_2 to calculate Y_2. Besides, the number of sensors is utilized as Z_3 to calculate Y_3.

Now, these three indicators (Z_1, Z_2, Z_3) can be got from scoring/counting of developers/experts. For example, five experts were asked to evaluate the matching degree with user requirements of one design entropy. Experts can give a score as 0, 1, 2, 3, 4 points (where 0 point means completely mismatching, 1 point means a little matching, 2 point means qualified, 3 point means matching well, 4 point means perfectly matching). The average score given by five experts represent Z_1 of this system, and Y_1 of the system can be calculate by $Y_1 = Z_1 * 0.25$.

For another example, the development team output five design entropies for a smart travel service system. The number of sensors of these five plans are 8, 4, 7, 9, 10, respectively. In other words, Z_3 of the five plans are 8, 4, 7, 9, 10, respectively. At this time, Y_3 can be calculated by the following formula:

$$Y_3 = \frac{Z_3 - min}{max - min} \qquad (4.9)$$

where max is the maximum value among Z_3 of these five design entropies (the maximum value is 10 in this example), and min is the minimum value (the minimum value is 4 in this example). Through the above formula, Y_3 of the five plans can be calculated respectively (e.g., Z_3 of plan 1 is 8, Y_3 of plan 1 can be calculated by $Y_3 = \frac{8-4}{10-4} = 0.667$).

Nevertheless, the disadvantage of this method is that when the design entropy is changed or a new design entropy is added to make the number of sensors beyond the current interval [min, max], developers need to recalculate the y of all plans. Therefore, developers can specify the maximum and minimum values of the number of sensors before designing. Furthermore, other methods can be used to calculate Y_3, such as:

$$Y_3 = e^{-Z_3} \qquad (4.10)$$

Other equations can also be used to calculate Y_n, that must enable Z_n mapped to the interval [0, 1] reasonably. Besides, some intelligent methods will be proposed in future research to automatically obtain the above indicators (Z_1, Z_2, Z_3) by utilizing AI methods. Furthermore, coefficient A_n is a weighting factor to represent the importance of Y_n in the different case. A_n is given by the stakeholders for expressing the significance of Y_n to the system. Thus $0 \leq C_{innovative} \leq 1$ and $0 \leq Y_n \leq 1$, where

$$A_1 + A_2 + \cdots + A_n = 1 \tag{4.11}$$

4.2.4 Iterative design entropy

Fig. 4.4 illustrates the process of information conversion in the usage and iterative design stage. For example, a smart office service system receives a piece of new information that "it is snowing outdoors." Since the smart office is indoors and its users have no plans to go out at present, this information does not influence the system currently. It is a noise which should be deleted. If it is not noise, x needs to be converted into x'. Iterative design entropy consists of the sum of the design entropy of all nonnoise and unconverted information x:

$$D_{iterative} = \sum_{x \in \mathcal{X}} D(x) \tag{4.12}$$

where $\mathcal{X}$ is the set of all noiseless/noise-free and unconverted information, and information x belongs to the set $\mathcal{X}$. As depicted in Fig. 4.5, x_i converted into x' (i.e., x'_1 to x'_m), through m items of channels. The DE of information x_i is denoted as:

$$D(x_i) = -\sum_{m \in \mathcal{M}} W(x^i_m) \log C(x^i_m) \tag{4.13}$$

where $\mathcal{M}$ is the set of all channels between information x_i and information after conversion, and channel m belongs to the set $\mathcal{M}$. $W(x^i_m)$ is the weighting factor of channel m, representing the correlation of information

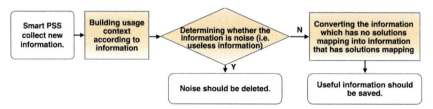

Figure 4.4 Process of information conversion in the usage and iterative design stage.

Design entropy theory   67

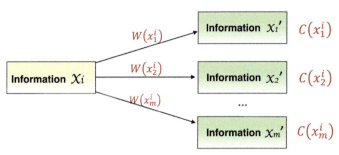

Figure 4.5 Conversion of information x_i.

x_i and x'_m. $W(x^i_m)$, which will be modified in a diverse context, is given by the stakeholders defined as:

$$W(x^i_1) + W(x^i_2) + \cdots + W(x^i_m) = 1 \qquad (4.14)$$

Additionally, $C(x^i_m)$ refers to the conversion ability to convert information x_i into x'_m through channel m, representing the single-channel conversion capability for a single piece of information in the system.

$$C(x^i_m) = A(x^i_m)_1 \times Y(x^i_m)_1 + A(x^i_m)_2 \times Y(x^i_m)_2 + \cdots + A(x^i_m)_n \\ \times Y(x^i_m)_n \qquad (4.15)$$

where $Y(x^i_m)_n$ is a parameter that represents the ability that, information x_i can be converted into information x'_m effectively, efficiently and economically. Since $0 \leq C_{innovative} \leq 1$, we let $0 \leq Y(x^i_m)_n \leq 1$. Each $Y(x^i_m)_n$ is determined by $Z(x^i_m)_n$, and different IC types n have different $Z(x^i_m)_n$. $Y(x^i_m)_1$ refers to the effective capability of completing the IC from x_i to x'_m, and is defined as:

$$Y(x^i_m)_1 = 0.01 \times Z(x^i_m)_1 \qquad (4.16)$$

where $Z(x^i_m)_1$ represents the user satisfaction degree (in a 100-point rating scale) for the IC from x_i to x'_m:

$$Z(x^i_m)_1 = \begin{cases} S(x^i_m)_1 - S(x^i_m)_2 + 100 & (S(x^i_m)_1 < S(x^i_m)_2) \\ 100 & (S(x^i_m)_1 \geq S(x^i_m)_2) \end{cases} \qquad (4.17)$$

where $S(x^i_m)_1$ and $S(x^i_m)_2$ respectively represents the average user satisfaction degree after and before this conversion. The improvement of average user satisfaction degree after this conversion is great, means the conversion ability of this channel is high.

Take a smart running service system as an example. The system collects a piece of information "the user has an error posture while running" and converts it into a piece of information "telling users his error running posture through voice prompt." Nevertheless, after the system giving voice prompts, many users are dissatisfied for the reason that they feel the system disturbing his sporting. The average user satisfaction degree after this conversion is lower than before: $S(x_m^i)_1 < S(x_m^i)_2$. The average satisfaction degree of users before IC $S(x_m^i)_2$ is 90, and the average satisfaction degree after conversion $S(x_m^i)_1$ is 85, then $Z(x_m^i)_1 = 95$, so $Y(x_m^i)_1 = 0.95$ can be obtained.

However, if the system converts the information "the user has an error posture while running" and into "telling users his error running posture through vibrating the smart bracelet once." In this case, most users are very dissatisfied with the vibrating prompt, and their satisfaction is even much lower than that of voice prompt. The average satisfaction degree after conversion $S(x_m^i)_1'$ is 60, then $Z(x_m^i)_1' = 70$, so $Y(x_m^i)_1' = 0.7$ can be obtained.

One can find that $Y(x_m^i)_1' < Y(x_m^i)_1$. It means that in this project, the effectiveness of converting information into "telling users his error running posture through voice prompt" is better than converting it into "telling users his error running posture through vibrating the smart bracelet once."

$Y(x_m^i)_2$ refers to the efficient capability of completing the IC from x_i to x_m':

$$Y(x_m^i)_2 = e^{-Z(x_m^i)_2} \tag{4.18}$$

where $Z(x_m^i)_2$ represents the total time of converting information x_i into x_m' and implementing the solution described by x_m', which is defined as

$$Z(x_m^i)_2 = T(x_m^i)_1 + T(x_m^i)_2 \tag{4.19}$$

where $T(x_m^i)_1$ refers to the converting time of determining the information x_m'. $T(x_m^i)_2$ refers to the average implementation time of realizing the solution described by x_m'. Take a smart office service system as an example. The system receives a piece of information "the user's eyes are very tired during office work." The system converts this information to the information "A message pops up on the computer interface to remind the user taking a break" in real-time according to the conversion plan already in the system. The converting time of this step is $T\left(x_m^i\right)_1 = 0$. After completing the IC, the user's computer receives the signal and pops up a prompt box in the upper right corner of the computer interface, which shows a reminder "you need to relax your eyes." The average implementation time of this step took 2 s: $T\left(x_m^i\right)_2 = 2$. And we can get $Z\left(x_m^i\right)_2 = T\left(x_m^i\right)_1 + T\left(x_m^i\right)_2 = 0 + 2 = 2$ and $Y\left(x_m^i\right)_2 = e^{-Z\left(x_m^i\right)_2} = 0.135$. It should be noted that for different types of information, their unit of the conversion time may be different, which needs to be set by the developers in advance. For example, if the service provided is maintenance service, the implementation time unit may be set to 1 day.

Then, $Y\left(x_m^i\right)_3$ refers to the economical capability to complete the IC from x_i to x_m':

$$Y\left(x_m^i\right)_3 = e^{-Z\left(x_m^i\right)_3} \tag{4.20}$$

where $Z\left(x_m^i\right)_3$ represents total costs associated with completing this conversion, and it is an indicator of the consumption of any economic resources. Therefore, $Z\left(x_m^i\right)_3$ is inversely proportional to $Y\left(x_m^i\right)_3$ and $C\left(x_m^i\right)$.

Following the previous example, the smart office service system receives a piece of information "the user's eyes are very tired during office work" and convert it to "reminding the employee taking a break." So the conversion cost is the operating expense required by the smart system provides "reminder rest service." However, if the system converted this information into "check and maintenance the computer screen of the users," then the conversion cost is the cost of manual services.

It should be noted that the unit of the conversion cost and time can be determined by the actual situation of each project. For the reason that there are no negative numbers for time and cost, Eqs. (4.18) and (4.20) can ensure that the value range of $Y\left(x_m^i\right)_n$ is [0, 1]. In fact, Eqs. (4.17), (4.18), and (4.20) can be changed to other appropriate equations according to the project, but it should ensure that after mapping $Z\left(x_m^i\right)_n$ to $Y\left(x_m^i\right)_n$, the value range of $Y\left(x_m^i\right)_n$ is [0, 1], and in a specific project, the equations should be uniform.

Additionally, Coefficient $A(x_m^i)_n$, which refers to the weighting factor of the importance of $Y(x_m^i)_n$ given by the stakeholders. Thus $0 \leq C(x_m^i) \leq 1$ and $0 \leq Y(x_m^i)_n \leq 1$, and

$$A(x_m^i)_1 + A(x_m^i)_2 + \ldots + A(x_m^i)_n = 1 \qquad (4.21)$$

4.3 Self-adaptable design process of smart PSS

Fig. 4.6 illustrates the self-adaptable design process of Smart PSS, which is enabled the DET. The design entropy of Smart PSS consists of three parts, including SCP design entropy, service design entropy, and the conversion plan.

DET gives explicit guidelines toward achieving innovation and iteration for Smart PSS. The stages of the systematic design methodology are as follows:

(1) *Developing the SCP and service design entropy.* In this stage, the development team should determine the target value of the innovative DE and the main functional demands of the system. The development team can design some initial design entropies for SCPs and e-services through brainstorming. Designers are suggested to express the design entropies

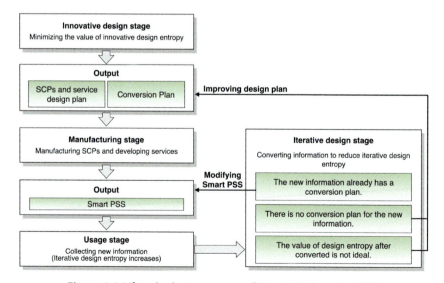

Figure 4.6 Lifecycle design process of Smart PSS by using DET.

by using blueprints or CAD models. Meanwhile, these plans must meet functional demands and under the target value of innovative DE. After that, the developers should calculate the value of the innovative DE of these initial design entropies. The plan with the lowest value should be chosen. The selected design entropy will be developed in-depth.

For example, a design team wants to develop a new smart water bottle service system. They first innovate four design entropies of the smart water bottle and its service. These design entropies with different shapes, colors, materials, modules, and e-services are expressed in the form of CAD models and blueprints. Nevertheless, how to choose the best one from these design entropies for the in-depth development? The team use DET. They calculate the innovative DE value of these four plans and find that the value of design entropy 2 is the lowest, so plan 2 is adopted as the final design entropy to develop in-depth.

(2) *Developing the conversion plan.* According to the final design entropy of SCP and e-service, Smart PSS information conversion plan can be output. The conversion plan is a part of the design entropy. The development team can use a conversion plan to predict and list the information x that will be received at a later stage. At the same time, the development team converts the design information x to the information x'. It needs to be emphasized that the design entropy mentioned in the DET is composed of the SCP and the service design entropy and the conversion plan. The blueprint or CAD model represents SCP and e-service design entropy, and the ICM outlines the conversion plan. Besides, it should be noted that these two plans are interrelated and interactive. When designing a new Smart PSS, the SCP and e-service design entropy should be formulated first. Meanwhile, the information conversion plan should be conceived based on the SCP and e-service design entropy. When iterating the design entropy, the conversion plan should be modified first, and the SCP and e-service design entropy can be updated if necessary.

Take the design team of smart water bottle service system as an example continuously. At this stage, developers begin to predict the usage contexts and the information that system might receive in different contexts, such as the information "the current temperature of water in the bottle is 30 degrees higher than this user's favorite drinking water temperature after exercise" in the context of user exercise, and the information "user has been working for 3 h without drinking water" in the context of user working in the office. Then the design team lists the converted information (e.g., "turn on the cooling function to cool the water to a suitable temperature" and "provide the service of reminding user drinking water") by utilizing the ICM.

(3) *Executing and iterating the conversion plan.* The system will convert the collected information according to the IC plan. At this time, the value of system iterative DE should be observed. Then when the value outweighs the standard value range, the development team should check and iterate the design entropy. It should be noted that the conversion plan should be customized according to the personal information.

Still take the smart water bottle service system as an example. During the usage stage, the system receives and converts a variety of new information, and monitors the iterative entropy of the system in real-time. At this time, the system receives a new information "this water bottle is too heavy" from the user's online review, and convert it to the information "replace the metal bottle cap module to a plastic bottle cap" according to the ICM. Some users replace the bottle cap module, but the time of purchasing a new module in the offline store is too long, the cost of the plastic cap is high, and the weight reduction effect after the replacement is not satisfactory. This led to a low value of the conversion ability of this IC, which resulted in a very high iterative DE value of this information. The design team quickly discovers this problem by monitoring the iterative DE of the system and immediately modified the design entropy.

4.4 Information conversion map tool

When the development team wants to build a conversion plan, some tools can be utilized, such as an ICM. The essential elements of the ICM are depicted in Fig. 4.7. Developers can mark the context in the dashed box, record the information x_i before the conversion in the yellow box, write the design entropy $D(x_i)$ of the information x_i in the white box, and mark the corresponding information x'_m after the conversion in the green box. Besides, the developer can note the weighting factor $W\left(x^i_m\right)$ above the line, and record the conversion ability $C\left(x^i_m\right)$ below the line.

Fig. 4.8 further depicts the core steps to build an ICM, which are elaborated below:

Step 1: Information collected in different usage contexts should be predicted and listed. Take a smart office service system as an example. In order to

Design entropy theory 73

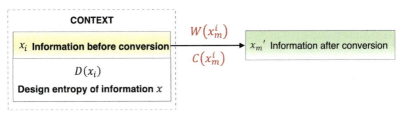

Figure 4.7 Essential elements of information conversion plan.

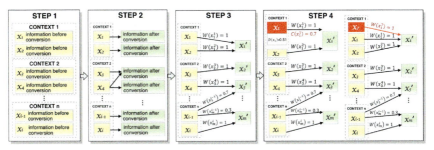

Figure 4.8 Core steps to develop the ICM of smart PSS.

construct the ICM, the design team first list the usage contexts of the system and the information that may be received in these contexts, e.g., in the context that employees are meeting in a conference room, the system will collect the information "an employee connects his computer with the projector to play the slideshow."

Step 2: The information after conversion should be planed, and the preliminary conversion plan should be organized. In the example of the smart office service system, the designer transformed the information "an employee connects his computer with the projector to play the slideshow" to the information "system automatically reduces the brightness of the lighting in the meeting room" and "system automatically turns on the microphone of the lecturer in the meeting room."

Step 3: The weighting factor and the information conversion plan should be marked and output. In the smart office service system example, designers set the weighting factor as 0.7 of the information conversion from "an employee connects his computer with the projector to play the slideshow" to the information "system automatically reduces the brightness of the lighting in the meeting room," and set the weighting factor as 0.3 of the information conversion from "an employee connects his computer with the projector to play the slideshow" to the information "system automatically turns on the microphone of the lecturer in the meeting room" based on their experience.

Step 4: The conversion plan during customers usage should be operated and iterated. In order to ensure that the value of iterative DE is within the expected range, the system should monitor the value of iterative DE in real-time when the clients use Smart PSS. This step summarizes the two situations when collecting a new piece of information, as follows:

(1) If the received information has been predicted in the conversion plan before, the system can directly activate the conversion channel of this information, and provide services, adjust parameters or replace modules according to the corresponding information x'_m. Then the system can obtain the iterative DE value of information x_1. If this value exceeds the standard, the developer can intervene and readjust the conversion plan. In the example, when the smart office service system receives the information "employees connect to the computer to play a slideshow" during the usage stage, the conversion channel of this information is activated. Then the system automatically reduces the brightness of the lighting and turns on the microphone of the lecturer in the meeting room. This conversion produces corresponding user satisfaction degree, time, and conversion cost. The system can calculate the iterative DE of this information. More information in the system is continuously collected and converted so that the iterative DE of the smart office service system can be obtained.

(2) If the newly collected information has not been predicted before, the DE of this information will tend to infinity. The system should modify the conversion plan, e.g., converting the information x_7 to an existing information x'_1. In the example, when the smart office service system collects a piece of new information without conversion plan during the usage stage, e.g., system collects the information "an employee suddenly falls on the ground." This information cannot be converted, resulting in its conversion ability $C(x^i_m)$ being 0, which leads to the iterative DE of this information tends to infinity, then it will inevitably make the iterative entropy of the entire system too high. At this time, the designer must immediately intervene to give a conversion plan to this information (e.g., this information can convert to the information "system alerts other employee in the office to find and help the people who fell on the ground").

4.5 Case study

This section provides a case study of smart travel assistant system (STA). Many elderly people want to travel to various regions after retirement. Most of them want to live in a local retirement home. An STA system should be developed to meet the functional needs of the elderly users (for example, the system should assist the elderly to adapt to the distinct places and send the health report of the elderly to their children).

In this project, the developers set the target innovative DE value to 0.7. Accordingly, the innovative design entropy of the initial design entropies must be equal to or less than 0.7. Fig. 4.9 shows the three initial design

Figure 4.9 An example of initial design entropies of the STA system.

entropies of the smart travel assistant system proposed by the development team through brainstorming. The design team calculated the value of the innovative design entropy of these three plans. Meanwhile, the design team used the matching degree with user requirements, adaptability measure, and the number of sensors as Y_1, Y_2, Y_3. According to the significance of Y_1, Y_2, Y_3 to the STA system, A_1, A_2, A_3 are recorded as 0.5, 0.3, and 0.2 respectively. We have:

$$D_{innovative1} = -\log(0.5 \times 0.8 + 0.3 \times 0.7 + 0.2 \times 0.4) = 0.535$$
$$D_{innovative2} = -\log(0.5 \times 0.6 + 0.3 \times 0.6 + 0.2 \times 0.8) = 0.644$$
$$D_{innovative3} = -\log(0.5 \times 0.7 + 0.3 \times 0.5 + 0.2 \times 0.6) = 0.690$$

After calculating and comparison, Plan A with the lowest innovative design entropy should be selected as for further in-depth development.

According to the design entropy A of the smart travel assistance system, the development team described the information conversion plan by means of an ICM. Fig. 4.10 summarizes the information that the smart travel assistance system will collect in the three primary usage contexts. Then, the development team listed the converted information, as shown in Fig. 4.11. Finally, as shown in Fig. 4.12, the development team marked the weighting factors and output the information conversion plan of the smart travel assistant system.

During the usage phase, information x_1 is collected. The system automatically activates the x_1 IC channel. Besides, the maintenance service described in x_1' is provided to the user. The conversion ability value and DE value of information x_1 are recorded (see Fig. 4.13). Because the DE of the information x_1 is within the standard range, designers did not intervene. Furthermore, as shown in Fig. 4.14, the system has collected new information x_{11} without a conversion plan. Then, the developers supplemented the conversion plan to convert x_{11} into the existing information x_1'. At the same time, as shown in Fig. 4.15, the system obtains another new information x_{12}, and the developers convert it into new information x_{11}'.

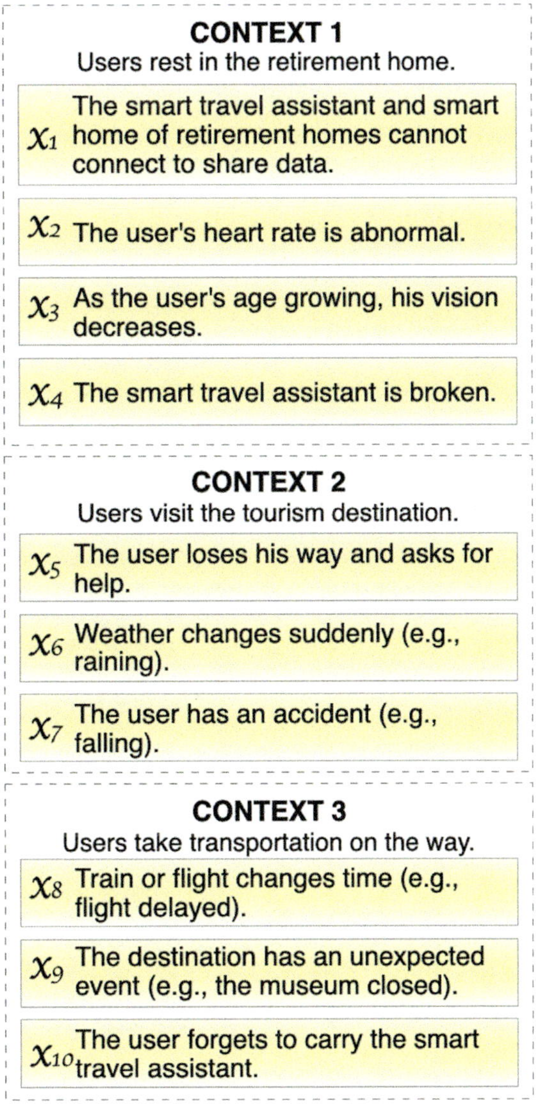

Design entropy theory   77

CONTEXT 1
Users rest in the retirement home.

X_1 The smart travel assistant and smart home of retirement homes cannot connect to share data.

X_2 The user's heart rate is abnormal.

X_3 As the user's age growing, his vision decreases.

X_4 The smart travel assistant is broken.

CONTEXT 2
Users visit the tourism destination.

X_5 The user loses his way and asks for help.

X_6 Weather changes suddenly (e.g., raining).

X_7 The user has an accident (e.g., falling).

CONTEXT 3
Users take transportation on the way.

X_8 Train or flight changes time (e.g., flight delayed).

X_9 The destination has an unexpected event (e.g., the museum closed).

X_{10} The user forgets to carry the smart travel assistant.

Figure 4.10 Information prediction list of the STA.

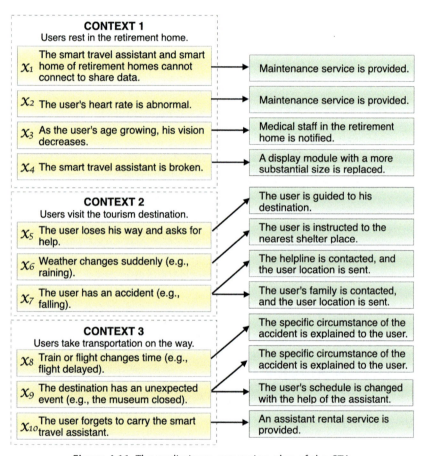

Figure 4.11 The preliminary conversion plan of the STA.

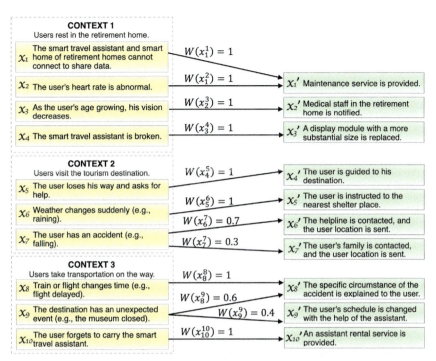

Figure 4.12 Output of the final conversion plan of the STA.

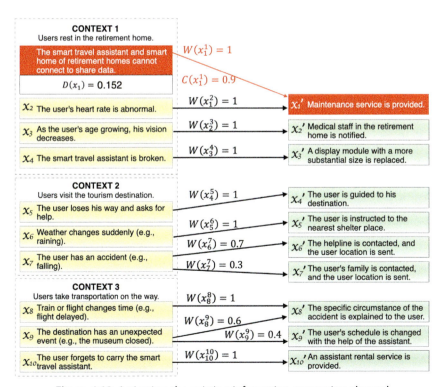

Figure 4.13 Activating the existing information conversion channel.

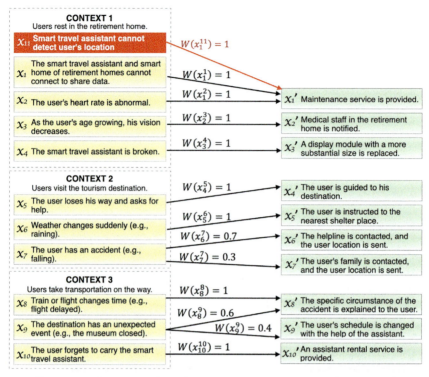

Figure 4.14 Converting the information x_{11} to an existing information x'_1.

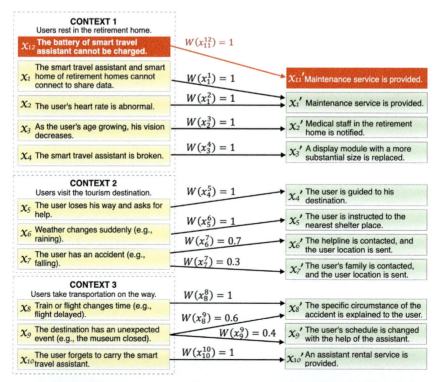

Figure 4.15 Converting the information x_{12} to a new information x'_{11}.

4.6 Summary

From the above content, it appears that DET can be promisingly utilized in Smart PSS development to satisfy the closed-loop design request, which is represented as the dynamic change of information and entropy in a balanced system. Design entropy theory focuses on not only innovative design phase but also iterative design/inverse design phase, which is critical to extend the lifecycle in Smart PSS. Meanwhile, it observes the status of real-time information conversion by monitoring iterative design entropy.

Moreover, the principles of DET can also be leveraged to guide the value co-creation and design with context-awareness issues as follows:

The value co-creation process can be described as the exchange of collecting information through the container. Furthermore, DET presents that stakeholders can set parameters which measure conversion ability together. Meanwhile, the user satisfaction degree of information conversion and the matching degree between design solutions and user requirements have been concerned to ensure users participation.

The design process with context-awareness denotes the process of reducing design entropy. DET suggests that Smart PSS should adjust its design/system to reduce design entropy in an information-driven manner triggered by the specific context. Moreover, it underlines the fact that collected information of Smart PSS from both hardware sensors and social sensors, should be utilized in perceiving and building the usage context.

In future works, AI techniques can be utilized to automatically obtain critical indicators in the usage stage, e.g., the user's physiological data, behavioral data, and environmental data can be collected to calculate the users' satisfaction of the information conversion with no direct human control. The iterative design entropy of the system can be obtained and monitored in real-time with less developer intervention. Meanwhile, a personalized ICM should be established in the Smart PSS for each user to match their requirements, and configure the product and service for users according to their individual map. In addition, an information conversion plan database can be established in the future.

References

Anderson, N. S., Norman, D. A., & Draper, S. W. (1988). User centered system design: New perspectives on human-computer interaction. *The American Journal of Psychology, 101*(1), 148.

Augusto, J., Kramer, D., Alegre, U., Covaci, A., & Santokhee, A. (2017). The user-centred intelligent environments development process as a guide to co-create smart technology for people with special needs. *Universal Access in the Information Society, 17*, 115–130. https://doi.org/10.1007/s10209-016-0514-8.

Choi, I. K., Kim, W. S., Lee, D., & Kwon, D. S. (2015). A weighted QFD-based usability evaluation method for elderly in smart cars. *International Journal of Human–Computer Interaction, 31*, 703–716. https://doi.org/10.1080/10447318.2015.1070553.

Chou, C. J., Chen, C. W., & Conley, C. (2012). A systematic approach to generate service model for sustainability. *Journal of Cleaner Production, 29–30*, 173–187.

Cong, J. C., Chen, C. H., & Zheng, P. (2020). Design entropy theory: A new design methodology for smart PSS development. *Advanced Engineering Informatics*.

Cong, J. C., Chen, C. -H., Zheng, P., Li, X., & Wang, Z. (2020). A holistic relook at engineering design methodologies for smart product-service systems development. *Journal of Cleaner Production, 272*(June), 122737.

Filho, M. F., Liao, Y., Loures, E. R., & Canciglieri, O. (2017). Self-aware smart products: Systematic literature review, conceptual design and prototype implementation. *Procedia Manufacturing, 11*(June), 1471–1480.

Gero, J. S. (1990). Design prototypes. A knowledge representation schema for design. *AI Magazine, 11*(4), 26–36.

Gero, J. S., & Kannengiesser, U. (2004). The situated function-behaviour-structure framework. *Design Studies, 25*(4), 373–391.

Govers, C. P. M. (2001). QFD not just a tool but a way of quality management. *International Journal of Production Economics, 69*(2), 151–159.

Gu, P., Hashemian, M., & Nee, A. Y. C. (2004). Adaptable design. *CIRP Annals - Manufacturing Technology, 53*(2), 539–557.

Gupta, R. K., Belkadi, F., Buergy, C., Bitte, F., Da Cunha, C., Buergin, J., … Bernard, A. (2018). Gathering, evaluating and managing customer feedback during aircraft production. *Computers and Industrial Engineering, 115*(December 2017), 559–572.

Gu, P., Xue, D., & Nee, A. Y. C. (2009). Adaptable design: Concepts, methods, and applications. *Proceedings of the Institution of Mechanical Engineers - Part B: Journal of Engineering Manufacture, 223*(11), 1367–1387. https://doi.org/10.1243/09544054JEM1387. http://journals.sagepub.com.

Hu, C., Peng, Q., & Gu, P. (2015). Adaptable interface design for open-architecture products. *Computer-aided Design and Applications, 12*, 156–165. https://doi.org/10.1080/16864360.2014.962428.

Ilevbare, I. M., Probert, D., & Phaal, R. (2013). A review of TRIZ, and its benefits and challenges in practice. *Technovation, 33*(2–3), 30–37.

Jindo, T., & Hirasago, K. (1997). Application studies to car interior of Kansei engineering. *International Journal of Industrial Ergonomics, 19*(2), 105–114.

Karat, J., & Watson, I. B. M. T. J. (1996). User centered design: Quality or quackery? *Conference on Human Factors in Computing Systems, 10*(4), 19–20.

Kim, J. W., Sul, S. H., & Choi, J. B. (2018). Development of user customized smart keyboard using smart product design-finite element analysis process in the internet of things. *ISA Transacions, 81*, 231–243. https://doi.org/10.1016/j.isatra.2018.05.010.

Köksal, Gülser, & Eğitman, Alpay (1998). Planning and design of industrial engineering education quality. *Computers and Industrial Engineering, 35*(3–4), 639–642.

Koswatte, K. R. C., Paik, I., Park, W., & Kumara, B. T. G. S. (2015). Innovative product design using metaontology with semantic TRIZ. *International Journal of Information Retrieval Research, 5*, 43–65. https://doi.org/10.4018/ijirr.2015040103.

Lee, C. H., Chen, C. H., & Trappey, A. J. C. (2019). A structural service innovation approach for designing smart product service systems: Case study of smart beauty service. *Advanced Engineering Informatics, 40*, 154–167. https://doi.org/10.1016/j.aei.2019.04.006.

Lee, C. H., Wang, Y. H., & Trappey, A. J. C. (2015). Service design for intelligent parking based on theory of inventive problem solving and service blueprint. *Advanced Engineering Informatics, 29*, 295–306. https://doi.org/10.1016/j.aei.2014.10.002.

Li, M., Wang, L., & Wu, M. (2014). An integrated methodology for robustness analysis in feature fatigue problem. *International Journal of Production Research, 52*, 5985–5996. https://doi.org/10.1080/00207543.2014.895443.

Li, Z., Tian, Z. G., Wang, J. W., Wang, W. M., & Huang, G. Q. (2018). Dynamic mapping of design elements and affective responses: a machine learning based method for affective design. *Journal of Engineering Design, 29*, 358–380. https://doi.org/10.1080/09544828.2018.1471671.

Liu, Z., Ming, X., Qiu, S., Qu, Y., & Zhang, X. (2020). A framework with hybrid approach to analyse system requirements of smart PSS toward customer needs and co-creative value propositions. *Computers and Industrial Engineering*.

Liu, Z., Ming, X., & Song, W. (2019). A framework integrating interval-valued hesitant fuzzy DEMATEL method to capture and evaluate co-creative value propositions for smart PSS. *Journal of Cleaner Production, 215*, 611–625.

Liu, A., Teo, I., Chen, D., Lu, S., Wuest, T., Zhang, Z., Tao, F., Cheng, J., Qi, Q., Zhang, M., Zhang, H., & Sui, F. (2019). Biologically inspired design of context-a ware smart products. *Engineering, 5*, 637–645. https://doi.org/10.1016/j.eng.2019.06.005.

Moehrle, M. G. (2010). MorphoTRIZ - solving technical problems with a demand for multi- smart solutions. *Creativity and Innovation Management, 19*, 373–384. https://doi.org/10.1111/j.1467-8691.2010.00582.x.

Nagamachi, M. (1995). Kansei engineering: A new consumer-oriented technology for product development. *International Journal of Industrial Ergonomics, 15*, 3–11.

Nagamachi, M., & Imada, A. S. (1995). Kansei engineering: An ergonomic technology for product development. *International Journal of Industrial Ergonomics, 15*(1), 1.

Nugroho, M. B. (2012). User experience innovation: User centered design that works. *Journal of Chemical Information and Modeling, 53*.

Peng, Q., Liu, Y., Gu, P., & Fan, Z. (2013). Development of an open-architecture electric vehicle using adaptable design. In A. Azevedo (Ed.), *Advances in sustainable and competitive manufacturing systems*. Heidelberg: Springer International Publishing. https://doi.org/10.1007/978-3-319-00557-7_7.

Qin, J., Zhang, W., Zhou, M., Chen, Z., Jin, Z., & Hao, Z. (2019). Interaction design of the family agent based on the CMR-FBS model (In Chinese with English abstract). *Packaging Engineering, 40*, 108–114. https://www.cnki.net/kcms/doi/10.19554/j.cnki.1001-3563.2019.16.016.html.

Rauch, E., Dallasega, P., & Matt, D. T. (2016). The way from lean product development (LPD) to smart product development (SPD), In: 26th CIRP Design Conference, 50, 26–31. https://doi.org/10.1016/j.procir.2016.05.081.

Riel, A., Kreiner, C., Messnarz, R., & Much, A. (2018). An architectural approach to the integration of safety and security requirements in smart products and systems design. *CIRP Annals, 67*, 173–176. https://doi.org/10.1016/j.cirp.2018.04.022.

Savarino, P., Abramovici, M., Göbel, J. C., & Gebus, P. (2018). Design for reconfiguration as fundamental aspect of smart products. *Procedia CIRP, 70*, 374–379. https://www.sciencedirect.com/science/article/pii/S2212827118300118.

Savransky, S. D. (2000). *Engineering of creativity: Introduction to TRIZ methodology of inventive problem solving*. CRC Press.

Shannon, C. E. (1949). A mathematical theory of communication. *Bell System Technical Journal, 27*(3), 379–423.

Shostack, G. L. (1982). How to design a service. *European Journal of Marketing, 16*(1), 49–63.

Sohn, S. C., Kim, K. W., & Lee, C. (2013). User requirement analysis and IT framework design for smart airports. *Wireless Personal Communications, 73*, 1601–1611. https://doi.org/10.1007/s11277-013-1269-7.

Suh, N. P. (1998). Axiomatic design theory for systems. *Research in Engineering Design, 10*(4), 189–209.

Suh, N. P. (2007). Ergonomics, axiomatic design and complexity theory. *Theoretical Issues in Ergonomics Science, 8*(2), 101–121.

Tao, F., Cheng, J., Qi, Q., Zhang, M., Zhang, H., & Sui, F. (2017). Digital twin-driven product design, manufacturing and service with big data. *The International Journal of Advanced Manufacturing Technology.* http://link.springer.com/10.1007/s00170-017-0233-1.

Targowski, A. (2005). From data to wisdom. *Dialogue and Universalism, 15*(5), 55–71.

Verdugo Cedeño, J. M., Papinniemi, J., Hannola, L., & Donoghue, I. D. M. (2018). Developing smart services by internet of things in manufacturing business. *DEStech Transactions on Engineering and Technology Research, 14*(icpr), 59–71.

Wodehouse, A. J., & Ion, W. J. (2010). Information use in conceptual design: Existing taxonomies and new approaches. *International Journal of Design, 4*(3), 53–65.

Wang, C. H. (2015). Using the theory of inventive problem solving to brainstorm innovative ideas for assessing varieties of phone-cameras. *Computers and Industrial Engineering, 85*, 227–234. https://doi.org/10.1016/j.cie.2015.04.003.

Wang, C. H., & Chin, H. T. (2017). Integrating affective features with engineering features to seek the optimal product varieties with respect to the niche segments. *Advanced Engineering Informatics, 33*, 350–359. https://doi.org/10.1016/j.aei.2016.10.002.

Wang, C.-H. (2017). Incorporating the concept of systematic innovation into quality function deployment for developing multi-functional smart phones. *Computers & Industrial Engineering, 107*, 367–375. https://doi.org/10.1016/j.cie.2016.07.005.

Wang, F., Li, H., & Liu, A. (2015). A novel method for determining the key customer requirements and innovation goals in customer collaborative product innovation. *Journal of Intelligent Manufacturing, 29*, 211–225. https://doi.org/10.1007/s10845-015-1102-0.

Wang, Y. H., Lee, C. H., & Trappey, A. J. C. (2017). Modularized design-oriented systematic inventive thinking approach supporting collaborative service innovations. *Advanced Engineering Informatics, 33*, 300–313. https://doi.org/10.1016/j.aei.2016.11.006.

Zairi, M., & Youssef, M. A. (1995). Quality function deployment: A main pillar for successful total quality management and product development. *International Journal of Quality & Reliability Management, 12*(6), 9–23.

Xu, Q. (2020). Optimal design of smarter tourism user experience driving by service design. In *AHFE 2019 International conferences on usability & user experience, and human factors and assistive technology*, Washington D.C., USA, *972* pp. 542–551). https://doi.org/10.1007/978-3-030-19135-1_53.

Zhang, J., Xue, D., & Gu, P. (2015). Adaptable design of open architecture products with robust performance. *Journal of Engineering, Design, 26*, 1–23. https://doi.org/10.1080/09544828.2015.1012055.

Zheng, P., Xu, X., & Chen, C. H. (2018). A data-driven cyber-physical approach for personalised smart, connected product co-development in a cloud-based environment. *Journal of Intelligent Manufacturing, 31*, 3–18. https://doi.org/10.1007/s10845-018-1430-y.

Zheng, P., Xu, X., Yu, S., & Liu, C. (2017). Personalized product configuration framework in an adaptable open architecture product platform. *Journal of Manufacturing Systems, 43*, 422–435. https://doi.org/10.1016/j.jmsy.2017.03.010.

Zheng, P., Wang, Z., Chen, C.-H., & Khoo, L. P. (2019). A survey of smart product-service systems: Key aspects, challenges and future perspectives. *Advanced Engineering Informatics, 42*(April), 100973.

CHAPTER 5

New IT-driven value co-creation mechanism

Contents

5.1 Value co-creation mechanism		87
5.1.1 Design theory perspective		87
5.1.2 IT-driven value co-creation toolkits		88
5.1.3 Smart, connected open architecture product		92
5.1.4 SCOAP and its service modeling for value co-creation		94
5.1.4.1 Modular design of SCP		*96*
5.1.4.2 Scalable design of SCP		*97*
5.2 Hybrid intelligence via crowd-sensing		98
5.2.1 Fundamentals		98
5.2.2 Generic framework		100
5.2.3 Incentive mechanism		102
5.2.4 Data collection and fusion		103
5.2.5 Cost-driven decision making for value generation		107
5.3 Case study		109
5.4 Chapter summary		112
References		113

The rapid development of information and communication technology (ICT) has driven products to evolve from usual products (i.e., mechanical or physical products) to smart products (i.e., usual products with embedded system), or information technology (IT)-enabled products (i.e., usual products with connectedness), and are converging to today's so-called smart, connected products (SCP) (i.e., usual products with embedded IT) (Porter & Heppelmann, 2014), as shown in Fig. 5.1.

In today's digital servitization era, innovation usually does not occur in the internal research and development (R&D) departments, but rather at the point of customer contact, where value is generated through co-creation (Parida, Sjödin, & Reim, 2019). However, most industrial firms, especially those operating a business-to-business (B2B) model are not ready for value cocreative innovation. Therefore, despite investing the large effort to develop digital services, they struggled to create real customer value with high revenues (Sjödin, Parida, Kohtamäki, & Wincent, 2020).

Smart Product-Service Systems
ISBN 978-0-323-85247-0
https://doi.org/10.1016/B978-0-323-85247-0.00010-4

© 2021 Elsevier Inc.
All rights reserved.

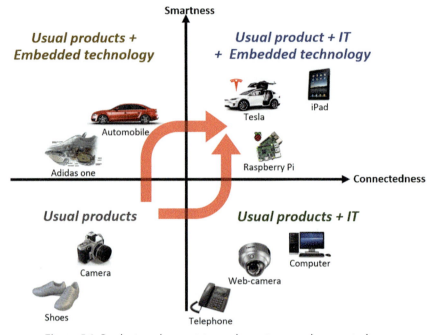

Figure 5.1 Product evolvement toward smartness and connectedness.

SCP, with embedded ICT components, have the abilities to collect, process, produce information, and somehow to "think by itself" (Rijsdijk & Hultink, 2009). For example, with a specific APP installed, a smart phone can be utilized for monitoring walking steps or informing upcoming events other than sending text message or ringing someone. Owing to the embedded open toolkits (Franke & Piller, 2004), SCP provides a promising way for mass personalization (Tseng, Jiao, & Wang, 2010). Users instead of designers can create their own products/services through adaptable hardware interface or application programming interface (API) in a value co-creation manner, following a specific set of rules or guidance. For instance, users can make new applications, such as a temperature monitor based on Raspberry Pi, to generate customized solutions by virtual prototyping in an online product platform. Conventionally, the detailed design information is either acquired via market research or assumed via professional knowledge, which is not effective and often results in an ever-rising new product development flop rates (Piller, Ihl, & Vossen, 2010). However, for SCPs, due to their connectivity, e.g., Internet of Things (IoT), companies have real–time access to the massive user-generated data, which enables the data-driven continuous design improvement (or "evergreen design") and next-generation product-service prediction (Porter & Heppelmann, 2015).

Since the SCPs serve as the tool and media to interact with customers/users and the foundation for value generation, hence, this chapter starts from an IT-driven perspective to understand how the development of SCPs and its services in product-service level (Chapter 5.1), and the overall Smart PSS deployment in system level (Chapter 5.2), can enable such value co-creation process.

5.1 Value co-creation mechanism

Cocreation is defined as an active, creative, and social collaborative process between users and manufacturer, aiming to creating values for customers (Piller, Ihl, & Vossen, 2010).

5.1.1 Design theory perspective

In literature, several existing prescriptive design theories can be adopted/adapted as the guideline for co-creation support to enable product personalization for individual customer satisfaction.

Two-stage based product design process, i.e., modular design and scalable design process (Simpson, Maier, & Mistree, 2001) or set-based design principles (Singer,Doerry, & Buckley, n.d.), for lean product development to reduce waste and continuously improve the product development processes (Tortorella, Marodin, Fettermann, & Fogliatto, 2016) by narrow down the dynamic changing and high-variety customer requirements in a systematic manner. With modular design at the macrolevel, and scalable design at the microlevel, customers' co-creation process can be guided to reduce the cycles of design iterations, and hence operate in a cost-effective manner. For example, an engineer-to-order product configuration system (Zheng, Xu, Yu, & Liu, 2017) was built based on this methodology.

Design with built-in-flexibility or *embedded open toolkit* (Franke & Piller, 2004), is a scalable attempt toward "open hardware" to postpone the flexibility of product differentiations into the customer domain after manufacturing rather than in the design stage (Piller, Ihl, & Steiner, 2010). Flexible knowledge and rules are predefined and embedded by designers to allow real-time modification along its life-cycle via adaptable interface. In such a way, users are enabled to adapt a physical product directly based on their own needs. A good application of this theory is the Adidas One shoes, which provide different cushion modes based on the user's behavior and road conditions.

Adaptable design (Gu, Xue, & Nee, 2009) for co-creation product changeability and life-cycle concerns. It provides two categories (i.e., design adaptability and product adaptability) with design principles (e.g., modular design, adaptable interfaces design) to systematically enable design flexibility under certain constraints in the early product development stage, and the easy replacing/upgrade of existing product modules in the later product usage or reconfiguration stage based on the ever-changing customer requirements. For instance, Xue, Hua, Mehrad, and Gu (2012) applied the adaptable design approach for creating the changeable wind power generator considering the whole product life cycle.

C-K theory (Hatchuel & Weil, 2002) for co-design iteration process. As a new form of product design theory, C-K theory goes beyond the pragmatic views of design iteration as a dynamic mapping process between the function domain and physical domain but modeled as an interplay between the *concept* (C) and *knowledge* (K) domains. Through the four operations, i.e., C-K, C-C, K-K, and K-C (Hatchuel & Weil, 2009), both the concept and knowledge are expanded iteratively, which can somehow represent the user-solution and design-solution space in the co-creation process for openness. For example, a novel function generated (e.g., an app for breathing condition detection) by customer based on the original product which is beyond the predefined scope of design solution.

Despite their advantages, each method has its own restricted scope of application stages, e.g., *embedded open toolkit* in the customer domain. Moreover, no approach alone directly supports the IT-driven co-creation process for open innovation in a smart, connected environment. Therefore, an appropriate SCP development approach with its co-creation context should be presented, which is depicted in the following sections.

5.1.2 IT-driven value co-creation toolkits

Several IT-driven value co-creation toolkits have been proposed and developed in literature. Generally, they can be classified into two categories, i.e., product configuration toolkit (Franke & Piller, 2004) and embedded open toolkit (Piller, Ihl, & Steiner, 2010).

Configuration toolkit, also referred to as product configuration system (PCS) or mass customization toolkit (Zheng et al., 2017), is a knowledge-based system to tailor a product according to the specific needs of a customer with a shorter lead time to market. It consists of a set of pre-defined attributes with constraints (rules) for customer to select within the product family scope. In the configuration process, the input is the customer's selection of existing attributes and the output is the recommended or target product derived from the system to fulfill customer requirements (CRs). In such a way, it bridges the gap between CRs and the end-product by only a series of attribute selection processes in a "configure-to-order" (CTO) model (Simpson et al., 2001). Moreover, it benefits the company by reusing existing design elements to provide customer-perceived product variety in the product family (Simpson, Jiao, Siddique, & Hölttä-Otto, 2014).

Embedded open toolkit, also known as product with built-in flexibility, was first proposed by Franke and Piller (2004) as a scalable attempt toward "open hardware," i.e., extending the open innovation paradigm toward tangible products, by embedding knowledge and rules about possible product differentiations into the product (Piller, Ihl, & Steiner, 2010). Other than the configuration toolkit, which only allows users to select or combine design modules and parameters in the design stage, the open toolkit is embedded in the product architecture, allowing real-time modification during the usage stage along its life-cycle via adaptable interfaces. It is claimed as a postponement method in new product development to increase design flexibility (Gross & Antons, 2009), which is built on shifting some specifications of the product into the customer domain after the product manufactured (Piller & Walcher, 2006).

Nevertheless, with the arrival of SCP, the concept of embedded open toolkit is broadened to the integration of open software innovation (e.g., apps in smart phones) and open hardware innovation (e.g., Raspberry Pi) as a whole, which is known as smart product with built-in flexibility. Due to the ever-increasing flexibility, it has the unique characteristics that users, instead of manufacturers, endorse the role of designing their own uses under certain guidance (Bénade, Brun, Le Masson, & Weil, 2016).

Fig. 5.2 depicts the evolvement of IT-driven value co-creation toolkits in correspondence with the product evolvement paradigm depicted in Fig. 5.1. Originally, usual product development is enabled by offline marketing strategies (e.g., user feedback, focus group). With the development of embedded technology and IT, embedded open toolkits (e.g., smart sensors) and online software toolkits (e.g., PCS) have been widely utilized, respectively. The integration of ICT technologies eventually guarantees the

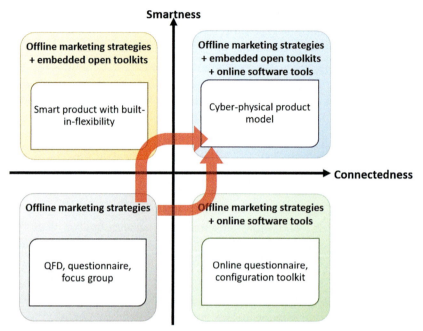

Figure 5.2 Evolvement of IT-driven value co-creation toolkits.

SCP with embedded smart open toolkits in a connected environment, where both online configuration toolkit and embedded open toolkits in an open environment for value co-creation process is achieved.

To achieve this, Smart PSS value co-creation is beyond the pragmatic view of systematic design theories (e.g., axiomatic design, function-behavior-structure design), as a dynamic mapping process between the function domain and physical domain, but modeled as an interplay between the *design solution space (DSS),* and the *user solution space (USS)* (Zheng, Lin, Chen, & Xu, 2018), as shown in Fig. 5.3.

Following this manner, the value co-creation process enabled by the development toolkits can be further explained as:

- *DSS → USS:* a manufacturer's defined solution space drives a user-generated solution, for example, a customized bicycle by undertaking online configuration.
- *USS → DSS:* a user-generated solution triggers the innovation of a manufacturer's design solution, for example, the real-time monitoring of machine tools usage by users motivating the new product.

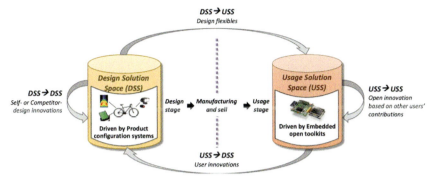

Figure 5.3 Integration of value co-creation toolkits for smart PSS development.

- *DSS → DSS*: a manufacturer's design solution urges self- or competitor-design innovation, for example, the introduction of fingerprint identification prompts a widespread application of smart phones.
- *USS → USS*: one or more user-generated solutions contribute to other users' innovations, for example, open source plugins contributed by one user can be edited by others, based on their own needs.

For PCS, it includes a set of predefined product attributes with configurable components (CCs), which users can tailor their preferred designs within the product family scope before manufacture, as shown in Fig. 5.4A. This type of codevelopment model is defined as *smart product with CCs*. The manufacturer rather than the user drives the design innovation, and users usually select from existing options by utilizing the online product configuration toolkit before it is manufactured (i.e., *DSS → USS*). Once determined, the end-product cannot be adjusted continuously in its life-cycle usage. For example, users can only change the color and type of the Fitbit wristband offered by the company.

For embedded open toolkit, it allows certain degree of customer design freedom by embedding knowledge and rules about possible product differentiations into the product after manufactured, as shown in Fig. 5.4B. This type of codevelopment model is defined as *smart product with built-in-flexibility*. The end-product can be adjusted continuously in its life-cycle usage. Nevertheless, it still correlates much with the original product offered by the manufacturer and user-generated data are often confined in the scalable usage options (i.e., *USS → DSS, DSS → USS*). For example, Adidas One is a kind of smart running shoe that users can change its modes, based on the usage context.

One can find that neither of them alone can fully allow user innovation along the whole product codevelopment process, and it is claimed that they should be integrated together, with an open environment (i.e., *DSS* → *DSS* and *USS* → *USS*) to achieve a higher degree of open hardware and software innovation flexibility.

5.1.3 Smart, connected open architecture product

Open architecture product (OAP) was first proposed by Koren, Hu, Gu, and Shpitalni (2013), defined as "one with a platform that allows the integration of modules from different sources in order to adapt product functionality exactly to the user's needs." Large companies, e.g., original equipment manufacturers (OEMs), tend to develop the common platform and define the interface. Small and medium-sized enterprises (SMEs), as third-party vendors, produce add-on modules that could be interfaced with the OAP platform (Koren, Shpitalni, Gu, & Hu, 2015). The customers engage in designing the options of their individualized product by using CAD software package developed by manufacturer or ordering certified modules from different vendors. Hu (2013) argued that an OAP platform that allowed for product compatibility/interchangeability of its functional features or components with standard mechanical, electrical, and information interfaces is the primary essential strategy of producing customer-oriented product for mass personalization.

Definition. Motivated by the concepts of OAP, and also to embrace the prevailing IT-driven value co-creation toolkits, a novel concept of smart, connected open architecture product (SCOAP) can be derived to enable value co-creation (Zheng, Xu, & Chen, 2018), which is defined as:

> *an IT-driven product consisting of physical, smart and connectivity components, with a platform containing both open hardware and software toolkits, that allows the integration of modules from different sources in order to adapt product and its service functionality exactly to the user's own needs throughout life-cycle.*

By this definition, SCOAP follows the adaptable design principles (Gu, Hashemian, & Nee, 2004), which aims to deliver designs and products fulfilling ever-changing CRs with an extended product life-cycle and represents nowadays product development toward highly smartness and connectedness. Moreover, it enlarges the scope of OAP as an open hardware innovation approach, by integrating open software innovation as well. Considering the open innovation environment, with many companies or suppliers joining the platform, each one can cooperate with many others by delivering or outsourcing part of its design resources or workload. Meanwhile, one can achieve abundant design resources and user information on demand, which again enables the "market-of-one."

***Scopes of* SCOAP**. According to the definition of SCOAP, smartness, connectedness, and openness are chosen as the three criteria to distinguish it with other types of products. Hence, the three-dimensional product category classification is depicted in Fig. 5.4A, where eight types of products can be derived. An example of each type is given in Table 5.1 as well, respectively. Furthermore, regarding the product evolution tendency in real-life, four typical kinds of products are elicited in Fig. 5.4B, where SCOAP only possess a small portion with limited real cases (e.g., a personalized iPad). This is not hard to understand, as there exists more constraints with the concerns of smartness, connectedness, and personalization. If one takes the other characteristics listed above into consideration (i.e., sustainability, data-driven manner, and servitization), the scope of SCOAP will again be narrowed. Another issue one should pay attention is that the openness has two folds in such paradigm, i.e., software openness (e.g., app store) and hardware openness (e.g., bus, slot, or hole). The overall performance is not a simple linear combination of them, but an integral consideration including all the dimensions. For example, the openness of a personalized Tesla is realized by the integration of various apps and different physical components into the product, while are enabled/constrained by the overall performance of its embedded smart components (e.g., microprocessing system) and connectivity components (e.g., WiFi module) as well.

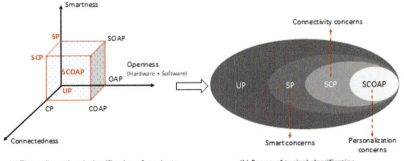

Figure 5.4 Classification of product categories based on smartness, connectedness, and openness.

Table 5.1 Typical examples of each product category.

Category	Type	Examples
Major tendency	Usual product (UP) Smart product (SP) SCP SCOAP	Shoes, doors, knives, etc. Automobile (general), robot arm, etc. Smart phones, iPaD, Tesla, etc. Personalized iPaD, Tesla, etc.
Other types	Smart open architecture product (SOAP) OAP Connected product (CP) Connected open architecture product (COAP)	Personalized automobile (general), etc. Lego bricks, etc. Wired telephone, water pipes, etc. Scalable hub, etc.

Nevertheless, in the IT-driven products, software can replace some hardware components or allow a single physical device to perform well at multilevels. For example, Apple iOS system upgrades without change hardware component to realize new functionalities (e.g., smart home). To depict the different levels of SCOAP, the isolated dots inside the cube in Fig. 5.4A are utilized to indicate the overall performance regarding the three criteria. One should be aware that the levels are only utilized to benchmark the proposed evaluation criteria, while companies should take overall consideration (e.g., cost, performance, positioning) to make value-added products in a more profitable manner.

Therefore, following the predefined interplay between USS and DSS, the value co-creation interactions between users and designers can be conducted either in the product prototype development process before the end-product is manufactured by utilizing PCS (Zheng et al., 2017), or after manufactured in the customer domain of usage with embedded open toolkits. Both spaces are expanded jointly through the action of design operators, viz. *design literation*, and the total space of design solution and user solution is defined as *product attainable solution space*, as shown in Fig. 5.5. In SCOAP, *USS* normally becomes much larger than *design solution space*. For instance, Raspberry Pi has been exploited in many applications by various users far beyond its original educational purpose.

5.1.4 SCOAP and its service modeling for value co-creation

As claimed by the authors before, in Smart PSS, both the digitalized services and e-services generated by the SCPs are in the form of software,

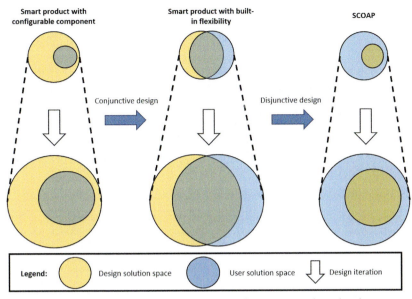

Figure 5.5 Toward smart, connected open architecture product development.

which can also be called software-as-a-service. Hence, to realize the SCOAP, a two-stage based product-service modeling method (i.e., modular design stage and scalable design stage) is proposed, as shown in Fig. 5.6. The modular design is at a macrolevel focusing on the performance of the entire product-service family design, while the scalable design is at a microlevel concentrating on the optimization of design specifications. Both guarantee the flexibility and openness of the SCPs in a bilevel manner.

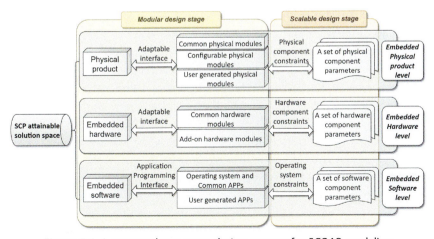

Figure 5.6 A proposed two-stage design process for SCOAP modeling.

Moreover, considering the unique characteristics of SCP, SCOAP modeling is classified into three levels: *physical product level, embedded hardware level,* and *embedded software level (service level),* respectively. Each subsequent level is modeled as an extension of the current one. For example, counting walking steps is an extended smart function application based on physical running shoes. Also, the degree of interdependency between the former level and the latter determines the type of data-driven codevelopment process, that is, *conjunctive design* or *disjunctive design.* For example, there lies a large interdependency between the physical product level, and embedded hardware/software levels of Adidas One running shoes, so that users cannot generate much innovation beyond the original SCP (*conjunctive design*). In contrast, iPhone, for instance, has no obvious interdependency between each level, which enables users to create many apps far beyond the basic function of calling service (*disjunctive design*). The detail of the proposed two-stage design process of three-level product-service modeling is described below.

5.1.4.1 Modular design of SCP

At the modular design stage, one should follow the principles of adaptable design (Gu et al., 2004), which stands for the ability of a design or a product to be adapted to new requirements and be reused when circumstances change, by adding or replacing certain modules through a predefined adaptable interface. Functional modeling is the key enabler for each design module to be independent of the others, so that the SCP can be easily upgraded or changed based on individual user needs. Meanwhile, the adaptable interfaces along with the design rules/constraints should be standardized by the platform provider to enable the user evaluation and design consistency by different stakeholders stepwise.

Physical product mainly consists of three types of modules: *common physical modules, configurable physical modules,* and *user-generated physical modules. Common physical modules* are the unchangeable physical modules, for example, the engine of an automobile. *Configurable physical modules* have a set of CCs, for example, gear box of an automobile. Both are predefined by the manufacturer. *User-generated physical modules* are add-on modules that can be designed by users individually through an adaptable interface, for example, a navigation module can be added to an automobile by inserting it into the slot.

Embedded hardware mainly consists of *common hardware modules* and *add-on hardware modules*. Normally, the hardware remains unchanged and sometimes can be replaced by upgrading the software. However, in order to be flexible enough for add-on functions, the hardware design should still consider *add-on hardware modules* enabled by adaptable interfaces (i.e., an open hardware toolkit). For example, the memory size of an Android device can be expanded by adding memory stick.

Embedded software mainly consists of *operating system with common APPs as the services*, that is, those preembedded in the hardware by the manufacturer, and *user-generated APPs* (e.g., online games), which are generated by customers or third-party developers, based on the API defined by the service providers. Software enables the realization of smart services and acts as the key value creation in the SCP. Its openness also determines the success of the final product. For example, an open-sourced Android operating system allows users to develop customized apps.

5.1.4.2 Scalable design of SCP

At the scalable design stage, the design specifications are "stretched" or "shrunk" based on a set of predefined rules (constraints) to satisfy individual user needs, which are fulfilled through parametric design optimization (Simpson, 2004). The optimization process is undertaken in the physical domain within the fixed product architecture defined by the modular design, such as specific features, specific values of parameter, and trade-offs between performance/solution parameters.

Physical product contains a set of *physical component parameters* corresponding to *configurable physical modules* and *user-generated physical modules*, following the *physical component constraints* predefined or approved by the manufacturer, respectively. For example, the diameter of wheel, as a CC, can be optimized within the predefined DSS.

Embedded hardware contains a set of *hardware component parameters* corresponding to the *add-on hardware modules*. The parameters should follow the *hardware component constraints* predefined by the manufacturer. For instance, the size of sensors should not exceed 2×2 cm for the add-on module in a PCB board.

Embedded software contains a set of *software component parameters* corresponding to the *user-generated software apps,* following the *operating system constraints*. Users can generate various APPs based on the specific programming language (e.g., JAVA for Android apps), libraries (e.g., jar in JAVA), versions (e.g., Java JDK version), etc.

Following this two-stage design modeling process at the product-service level, the open architecture and flexible Smart PSS for active user participation in co-creation process can be made. Nevertheless, in order to make the entire value co-creation process work, a systematic framework should be brought up at the system level as well.

5.2 Hybrid intelligence via crowd-sensing

On the other hand, in order to leverage the massive SCPs and to empower a large number of users to contribute their generated/sensed data for value co-creation of Smart PSS development, a systematic framework with proper mechanisms should be considered as well in the system level.

5.2.1 Fundamentals

Fig. 5.7 outlines the typical approach and the type of intelligence leveraged for value generated along the three PSS evolvement phases defined in Chapter 2.

In the *Internet-based PSS* phase (2000—), the major concern of IT-driven value creation lies in the efficient delivery of data/information with e-

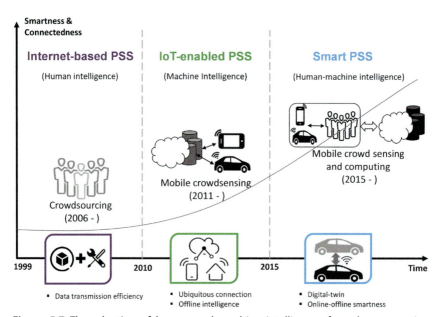

Figure 5.7 The adoption of human and machine intelligence for value generation among three PSS evolvement phases.

commerce service platform established. *Crowdsourcing* (Howe, 2006), as the practice of obtaining needed services or content by soliciting contributions from a crowd of people, especially from an online community, is a main way for user-generated design innovation.

Meanwhile, in the *IoT-enabled PSS* phase (2010—), empowered by the ubiquitous connectivity enabled by the IoT (Kevin, 2009), billions of end devices are connected to the Internet. In this phase, the major concern is the machine intelligence for value co-creation. Sensing data are collected and interchanged among the networked devices, which interact with real "things" such as sensors, actuators, and RFID, to realize value generation on the Internet with more intelligence. A typical example is the traffic route selection of Google Maps based on GPS in the mobile phones.

However, the emerging *Smart PSS phase* (2014—) is enabled by the prevailing adoption of SCPs, and the cutting-edge digital technologies (Zheng, Wang, Chen, & Khoo, 2019). SCP communicates with the others via the IoT infrastructure, where massive user-generated data and product-sensed data can be obtained through MSNs and heterogenous built-in sensors, adapted at a component and system level autonomously based on intelligent algorithms and big data analytics (Lee, Kao, & Yang, 2014).

In this phase, an emerging concept named mobile crowd sensing (MCS) was first coined by Ganti et al. (Ganti, Ye, & Lei, 2011) referring to a broad span of community sensing paradigms, with participatory sensing and opportunistic sensing at the two ends considering the level of user involvement. By leveraging large-scale mobile devices and empowering massive users to share surrounding information or accomplish specific sensing tasks, MCS has the advantages of high mobility, scalability, and cost effectiveness, which is superior to the static sensing infrastructures and can often replace them (Ma, Zhao, & Yuan, 2015). Owing to its great advantages, MCS has been widely adopted in many areas, such as traffic planning (Yang et al., 2017), unmanned vehicle control (Zhang et al., 2018), landmark measurement (Jordan et al., 2013), to name a few. Guo et al. (2015) further introduced a concept named mobile crowd-sensing and computing (MCSC) by taking both machine (e.g., sensing data) and human intelligence (e.g., crowdsourcing) in the MCS into an overall consideration. It extends the scope by leveraging both participatory sensory data from mobile devices (offline) and user-contributed data from MSN services (online). Hence, other than only collecting data from physical sensors, the MCS participants, acting as the "social sensors," have the ability to analyze data and transform into valuable knowledge with context-awareness (Xu et al., 2017). Owing to its significance, as the hybrid intelligence approach (Guo et al., 2015), MCS is adopted and further enhanced in the industrial environment to enable the Smart PSS value co-creation process.

5.2.2 Generic framework

To ensure Smart PSS development via MCS, a conceptual framework for manufacturer/service provider-user value co-creation is proposed, as shown in Fig. 5.8. It mainly consists of four layers, i.e., *physical resource layer, networked platform layer, service composition layer,* and *service application layer,* conducted in a platform-based data-driven manner.

Physical resource layer. It consists of a large scale of smart, connected devices, including both static sensing devices (e.g., 3D printers) and smart mobile devices (e.g., smart phones). Each device is assigned with a universal unique identifier (UUID) for easy identification and retrieval. They both serve as the main data collectors for data acquisition from different stakeholders participated in the sensing task in a wired (e.g., LAN) or wireless

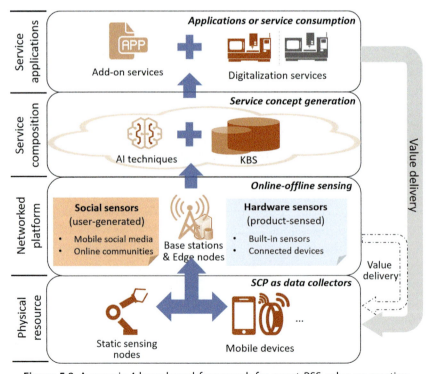

Figure 5.8 A generic 4-layer based framework for smart PSS value co-creation.

(e.g., WiFi or 4G/5G) environment. Meanwhile, other external data sources from the Internet (e.g., product information, service documents) should also be involved, as physical resources, to acquire the reliable domain-related knowledge. For cases, industrial devices (e.g., assembly line) can communicate with mobile devices (e.g., smart phones) through specific communication protocols so that information can be otherwise collected by the mobile devices alone as the wireless terminals.

Networked platform layer. A robust and cost-effective IT infrastructure is constructed in this layer and serve as the foundation for ubiquitous connectivity and should be provided to handle the SCP communication and data preprocessing. Massive user-generated online data (e.g., ratings, text feedback) from MSN and/or online communities as social sensors and product-sensed offline data (e.g., location, acceleration, pressure) from built-in sensors as hardware sensors are collected and preprocessed. In return, the service providers should give certain incentives to the effective contributors. The sensed wireless data can be submitted to data collectors via access to macrobase stations (MBSs) or submitted through little access points (LAPs), such as microbase stations and relay stations from nearby data collectors deployed by manufacturer/service providers (Ota, Dong, Gui, & Liu, 2018). The LAPs work as the middleware to not only receive data, but also preprocess it (e.g., filtering, cleaning) before submitting to the data collectors. Nevertheless, due to the limited sensing coverage of LAPs, large-scale mobile devices can also act as temporary relay stations for relaying data collected to the LAPs.

Service composition layer aims to manage encapsulated services based on individual request. It is mainly conducted in the advanced edge-cloud computing environment, where a proper knowledge modeling approach should be established by considering product-service-context information integrally. Hence, end devices are not only data consumers from the cloud, but also data producers performing computing offloading, data storage, caching and processing, as well as distribute request and delivery service in the edge plane to users (Shi, Cao, Zhang, Li, & Xu, 2016). Furthermore, an automatic manner should be established so that valuable knowledge can be extracted, stored, and further analyzed from the massive multi-source data collected for solution development purposes.

Service application layer emphasizes the value generation process, which includes both e-services and digitalized services for uses' applications (i.e., service consumers). In the Smart PSS context, these smart services as add-on values, are delivered to the customers by embedding them in the SCPs in

102 Smart Product-Service Systems

the IoT-enabled environment. Therefore, an effective approach should be proposed to trigger the solution recommendation/prediction in the context and manage the evolving solution design knowledge properly. Hence, the Smart PSS service applications can be conducted in a circular manner.

5.2.3 Incentive mechanism

On the other hand, the success of MCS-enabled Smart PSS value co-creation highly depends on the quantity of participants to guarantee the coverage and reliability (Jaimes, Vergara-Laurens, & Raij, 2015). Individuals may feel reluctant to participate and share their sensing knowledge due to the risk (e.g., data privacy) or cost (e.g., data transmission, energy consumption) raised thereafter. Hence, it is very critical to motivate users' engagement while maintaining the reliable data in a cost-effective manner. Based on the recent survey (Zhang, Yang, Sun, Liu, & Tang, 2016), most existing studies emphasizes the incentive mechanism issues, including: *form of rewards*, such as monetary (e.g., reverse auction (Lee & Hoh, 2010)), services (e.g., social welfare (Luo & Tham, 2012)), or entertainment approach (e.g., game (Jordan, Sheptykin, Grüter, & Vatterrott, 2013)); *target object*, such as customer-centric (e.g., user fairness (Yang, Xue, Fang, & Tang, 2012)) or platform-centric (e.g., service provider benefits (Faltings, Li, & Jurca, 2014)); and *level of participation*, such as opportunistic (e.g., urban sensing; Lane, Eisenman, Musolesi, Miluzzo, & Campbell, 2008) or participatory (e.g., route planning; Yang et al., 2017). Only recent works began to consider the dynamic changes of the overall sensing network including service providers, consumers, and data collectors (Ota et al., 2018) as a whole.

To motivate massive users' participation to elicit useful requirement information and to fulfill the sensing task, this work adopts a monetary approach based on the teaching cost (Zhang & Min, 2016), as depicted in Table 5.2.

Table 5.2 Teaching cost matrix.

User/service provider action	I	$\sim I$
a	λaa	λar
p	λpp	λpp
r	λra	λrr

Where *a, p, r* stand for the design action determined by the service provider, viz. accept, pending, and reject, respectively. Meanwhile, *I* or $\sim I$ stands for the user's action to conduct the design or not. $\lambda \in \{\lambda_{aa}, \lambda_{rr}, \lambda_{ar}, \lambda_{ra}, \lambda_{pp}\}$ stands for the cost/reward given to the users as incentives. Generally, only when user's action matches with service provider's, will a reward be given, viz. λ_{aa} or λ_{rr}. However, in the three-way based incentive model, the pending situations will result in a teaching cost, viz. λ_{pp} to acknowledge the users giving pending feedback, however, the reward will be much less than λ_{aa} or λ_{rr} since service providers need more accurate responses. Also, the misclassification will result in no rewards, i.e., λ_{ra} or λ_{ar}.

5.2.4 Data collection and fusion

Data collection. The heterogeneous data collected, including user-generated data, product-sensed data and other existing data sources contain various formats, including numerical data (e.g., distance, temperature) or nonnumerical data (e.g., text, audio, video). For the former one, to deal with the high variety of discrete numbers, specific ranges should be predefined by service providers to categorize them into different classes. For example, in Table 5.3, the distance by GPS varied from 1.65 to 58.09 km can be further classified into very short (VS) (<10 km), short (S) (10 −20 km), medium (M) (20 −50 km), and long (L) (50−100 km). Meanwhile, for the latter one, specific semantics should be extracted and again categorized based on the information fusion techniques. One may refer to (Guo, Tang, & Zhang, 2017) for more details. For example, the image can be extracted by the RGB value, x-axis dimension and y-axis dimension, and categorized into the set of {colored, black, and white}. Hence, the heterogeneous data should be fused into a consistent manner.

Information fusion. The design in the hybrid crowd-sensing environment can be structured as a 4-tuple information table:

$$T = (U, A, V, f) \tag{5.1}$$

where $U = \{x_1, x_2, \ldots, x_{|U|}\}$ is a nonempty finite set of design records, as the universe. $A = \{a_1, a_2, \ldots, a_{|A|}\}$ is a nonempty finite set of attributes and $\forall a \in \{M_H, M_S, S, DI\}$, where M_H is the set of MCS hardware sensing attributes (e.g., GPS location), M_S is the set of MCS social sensor data (e.g., user rating), S is the set of static sensing data (e.g., failure mode), as shown in Table 5.2. They together form the conditional attributes. While DI is the set of design decisions made by service providers, as the decision attributes.

Table 5.3 An example of water dispenser maintenance service records.

Objects		Condition set			Decision set
No.	Failure mode[a]	User rating	Distance by GPS (km)	Maintenance cost (SGD)	Service provider action
1	CM	5	3.17 (short)	13 (very low)	Yes
2	CM	5	1.65 (short)	19 (very low)	No
3	TDS	1	2.24 (short)	46 (medium)	Yes
4	HM	5	5.54 (short)	26 (low)	No
5	M	4	58.09 (long)	85 (high)	No
...	...	...	...	...	...
7043	E	2	4.27 (short)	22 (low)	Yes
7044	HM	1	15.19 (medium)	35 (low)	Yes
7045	FV	4	18.85 (medium)	9 (very low)	No

[a]Flow volume exceeded (FV); General electrical problem (E); Total dissolved solids > 60 (TDS); Heater malfunction (HM); General mechanical problem (M); Cooling malfunction (CM).

$A = M_H \cup M_S \cup S \cup DI$, and $M_H \cap M_S \cap S \cap DI = \varnothing$. V_a is a nonempty set of values for an attribute $a \in A$. $f: U \times A \rightarrow V$ is an information function, where $f(x_i, a_l) = v_{il}$ $(i = 1, 2, \ldots, |U|, l = 1, 2, \ldots, |A|)$ denotes the attribute value of object x_i under a_l.

Three-way decision theory [54] is an extension of decision-theoretic rough set approach based on the rough set theory [55] to deal with situations where three possible decisions exist. It has been widely adopted in various applications, such as movie recommendation, filtering spam email, to name a few. It has the unique advantages of scalability, i.e., computing the thresholds of boundary region with flexibility. Therefore, it can be adapted in the hybrid crowd-sensing environment for Smart PSS design, to enable information fusion in a structured manner.

In this study, after the data have been transformed into a consistent manner, the indiscernibility relation of the subset of attributes $A_S \in A$ can be defined as (Pawlak, 1982):

$$\text{IND}(A_S, V) = \{(x, y) \in A_{S^2} | \forall a \in V, f(x) = f(y)\} \tag{5.2}$$

where two objects x and y are indiscernible with respect to A_S if and only if they have the same value on every attribute in A_S, and for simplicity, the equivalence class of $x \in A_S$ is denoted by [x] in this work where

$$[x] = \{y \in A_S | (x, y) \in \text{IND}(A_S, V)\} \tag{5.3}$$

The partitioning of A_S induced by V is represented as:

$$A_S / V = \{s_1, s_2, \ldots, s_N\} \tag{5.4}$$

where $\forall s \in A_S / V$ is an equivalence class, and $\forall (s_i, s_j) \in A_S / V$, $s_i \cap s_j = \varnothing$, and hence, $P(s)$ represents the probability that a design action is needed (i.e., $f(y) = Y$) according to the condition attributes:

$$P(s) = |\{y \in A_S | f(y) = Y\}| / |A_S| \tag{5.5}$$

For example, the first two rows in Table 5.2 have the same set of condition attributes and values, while they result in different classes in the decision (1 Y, 1 N), hence, the probability of its service action is 50%. Then, the expected cost associated with taking different actions can be written as following equations:

$$C_{aa} = \lambda_{aa} P(x|I) + \lambda_{ar} P(x|\sim I) \tag{5.6}$$

$$C_{pp} = \lambda_{pp} P(x|I) + \lambda_{pp} P(x|\sim I) \tag{5.7}$$

$$C_{rr} = \lambda_{ra} P(x|I) + \lambda_{rr} P(x|\sim I) \tag{5.8}$$

where $P(x|I)$ is the conditional probability of the object x in condition I. According to Bayesian decision procedure, one can find the minimum–cost decision rules can be written as:

$$\text{If } C_{aa} < C_{pp} \text{ and } C_{aa} < C_{rr}, \text{ decide } x \in \text{Accept} \tag{5.9}$$

$$\text{If } C_{pp} < C_{aa} \text{ and } C_{pp} < C_{rr}, \text{ decide } x \in \text{Pending} \tag{5.10}$$

$$\text{If } C_{rr} < C_{aa} \text{ and } C_{rr} < C_{pp}, \text{ decide } x \in \text{Reject} \tag{5.11}$$

To simplify the rules and follow the incentive mechanism, some constraints are added:

$$P(x|I) + P(x|\sim I) = 1 \tag{5.12}$$

$$\lambda_{aa}, \lambda_{rr} \geq \lambda_{pp} > 0, \text{ and } \lambda_{ra} = \lambda_{ar} = 0 \tag{5.13}$$

$$0 < C_L \leq C_U < 1 \tag{5.14}$$

The C_L and C_U are the lower and upper thresholds of the pending region, where probabilities below C_L are in the reject region, ones above C_L are in the accept region, and ones in-between in the pending region, respectively. The threshold values of C_L and C_U can be further calculated as:

$$C_U = \frac{\lambda_{rr} - \lambda_{pp}}{(\lambda_{rr} - \lambda_{pp}) + (\lambda_{pp} - \lambda_{ra})} = \frac{\lambda_{rr} - \lambda_{pp}}{\lambda_{rr}} \tag{5.15}$$

$$C_L = \frac{\lambda_{pp} - \lambda_{ar}}{(\lambda_{pp} - \lambda_{ar}) + (\lambda_{aa} - \lambda_{pp})} = \frac{\lambda_{pp}}{\lambda_{aa}} \tag{5.16}$$

Hence, the equations can be denoted as:

$$\lambda_{rr} = \frac{1}{1 - C_U} = \lambda_{pp} \tag{5.17}$$

$$\lambda_{aa} = \frac{1}{C_L} = \lambda_{pp} \tag{5.18}$$

Finally, classification decision rules are obtained as:

$$\text{If } P(x|I) > C_U, \text{ decide } x \in \text{Accept} \tag{5.19}$$

$$\text{If } C_L \leq P(x|I) \leq C_U, \text{ decide } x \in \text{Pending} \tag{5.20}$$

$$\text{If } P(x|I) \leq C_L, \text{ decide } x \in \text{Reject} \tag{5.21}$$

5.2.5 Cost-driven decision making for value generation

From the aforementioned equations, one can obtain the total and average cost for user participation as follows:

$$T_c = \lambda_{aa}R_{aa} + \lambda_{pp}R_{pp} + \lambda_{rr}R_{rr} \qquad (5.22)$$

where R_{aa}, R_{rr} are the total numbers of users who provide the accurate feedback to design action in the accept region and reject region, respectively, and R_{pp} stands for the total numbers of users providing pending reviews.

To maintain the minimum active participation while not exceeding budget, random forest approach is adopted in this research, to predict DI and compute Tc and number of people rewarded based on the three-way incentive model. To test its performance, the total dataset is randomly divided into a training set and testing set to conduct the learning:

Step 1: Construction of random decision tree. In the training set, decision-tree learners build a tree by recursively partitioning the data, as depicted in Algorithm I. These trees are merged together to form a random forest.

ALGORITHM I. Construct random decision trees

Input: Training dataset (D_T), Condition attributes (A_C), Probability (P), Design action (DI)
Output:　　A Random Decision Tree (CNode)
Method:　　New Random Tree

```
 1  P = P(D_T); // probability of the training set D_T based on Eq. (5)
 2  DI = DI(P);
 3  CNode = NewRandomTree (D_T, A_C, DI);
 4  found = false; // Randomly select A_C and split (Line 6 - 13)
 5  For (a∈A_C)
 6       A_C = A_C − {a};
 7       If (InformationGain (a) > 0) then
 8             found =true;
 9             break;
10      End if
11  End for
12  If (not found) then
13        CNode.children = Null;
14        return CNode;
15  End if
16  CNode.splittingAttribute = a ; // Construct random tree (Line18 - 26)
17  Na = number of attribute values of a;
18  CNode.children = newbulidRandomTree [Na]
19  For (i = 1 to Na)
20        D_T (i) = {a∈ D_T | a (x) = i}
21        DI = DI (D_T (i));
22        CNode.children [i] = bulidRandomTree (D_T(i), A_C, DI);
23  End for
24  return CNode
```

108   Smart Product-Service Systems

Step 2: Design action result prediction. Each random decision-tree produces a prediction result, i.e., *P(s)*, based on the conditional/decision attributes and values. By leveraging theses probabilities, one can obtain the *P* in the three-way model, as shown in Algorithm II.

ALGORITHM II. Design action prediction based on random decision tree

Input: Current node (*CNode*), Test dataset (*D_t*)
Output: *DI* Prediction Result (*Pt*)
Method: PredictionbyRandomDecisionTree (PRDT)

1   a = CNode.splittingAttribute;
2    $j = a (D_t)$;
3   **If** (CNode.children = NULL) **then**
4   **return** CNode.*DI*;
5   **Else if** (CNode.children [j] = NULL) **then**
6   **return** CNode.*DI*;
7   **Else**
8   **return** PRDT (CNode.children [j], *D_t*);
9   **End if**

Step 3: Computation of total cost and number of people rewarded. The predefined thresholds C_L and C_U are leveraged to compare with the value *P* to classify different categories based on Eqs. (5.19)−(5.21). Then, both the total number of people rewarded and the cost can be calculated by the cost matrix and Eq. (5.22), as shown in Algorithm III.

ALGORITHM III. Computation of total cost and number of people rewarded

Input: Dataset of condition attributes and values (*D_a*), Thresholds (*C_U*, *C_L*), Cost for Pending (*C_p*), *DI* Prediction Result (*Pt*), Number of people rewarded (*Num*=0)
Output: Total Cost (*T_c*), *Num*

1   N = merge D_a and *Pt* together;
2   ClassofDesignInnovation(N) using Eq. (19-21);
3   **Define** an empty list L_1;
4   **For** ($i \in C_U$)
5   Calculate rewards using Eq. (17);
6   **Define** an empty list L_2;
7   **For** ($j \in C_L$):
8   Calculate awards using Eqs. (18);
9   **If** ClassofUserAction*(N_{ij})* == ClassofDesignInnovation*(N_{ij})*
10   **then** *Num* ++;
11   Calculate T_c using Eqs. (22);
12   add T_c to a list L_2;
13   **End for**
14   add list L_2 to list L_1;
15   **End for**
16   Convert list L_1 to dataframe;
17   **return** T_c, *Num* and corresponding C_U, C_L

5.3 Case study

To demonstrate the feasibility and advantage of the proposed approach, smart maintenance service design, as a typical kind of Smart PSS solution design is adopted, and a case study on a smart water dispenser product (SWD) made by company X is chosen. Unlike most existing companies undertaking maintenance services in a "on call"-based manner with service records documented manually, company X aims to obtain cost-effective reliable data sources from end users through an app and potentially automate the service recommendation process by leveraging existing datasets (i.e., service records). The end-product SWD, as the static sensing node, is equipped with several embedded sensors to detect water pressure (207–827 kPa), water temperature (0.6°–48.0°C), flow rate (1.39–1.89 Lpm), FV threshold (800 L), and total dissolved solids (TDS) (<40 etc. as the S). It can communicate with smart mobile phones with a specific app installed via Bluetooth module to monitor its real-time conditions for failure mode detection. Meanwhile, users can contribute their social sensing data (i.e., M_S), e.g., comments and ratings, and real-time MCS data, e.g., GPS and images (i.e., M_H), to the online community authorized by the manufacturer/service provider. Following such a manner, the SWD maintenance service can be performed in a user-centric and cost-effective manner with wide coverage, and the descriptive architecture of its hybrid crowd-sensing network is shown in Fig. 5.9.

For simplicity, as partially shown in Table 5.3, a total of 7045 service design action (i.e., DI) records (.csv file in the supplementary materials) of product A is analyzed in this research. The initial cost of λ_{pp} is set as 10 SGD, while the records are transformed into predefined categories as follows: *failure mode* (flow volume exceeded (FV); general electrical problem (E); TDS > 40; heater malfunction (HM); general mechanical problem (M); cooling malfunction (CM)); *user rating* (rating scale 1–5) ({1, 2} (not acceptable); 3 (pending); {4, 5} (acceptable)), *distance by GPS* (service zones) (very short (<10 km); short (10–25 km); medium (25–50 km); long (>50 km)). *Maintenance cost* (very low (<20 SGD); low (20–40 SGD); medium (40–60 SGD); high (60–100 SGD); very high (>100 SGD)). In this study, *failure mode* is calculated by the microprocessor in the existing SWD based on the abnormal signals detected from the sensors, *user-rating* is predefined in a five-point rating scale, and *distance by GPS* is measured based on the distance between the location of the mobile device and the actual service provider. The data analytics and visualization process are run in the *Jupyter* notebook web application written by Python 3 programming language. The authors exploit several existing libraries, such as *pandas* and

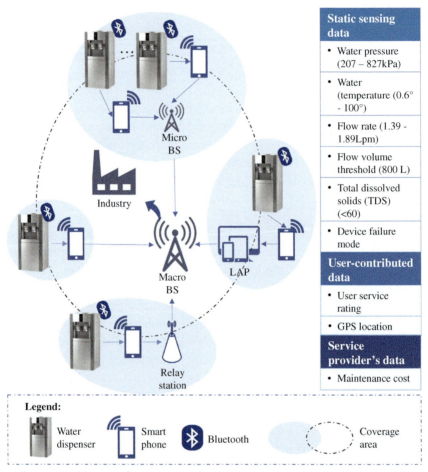

Figure 5.9 Diagram of smart water dispenser maintenance in an MCS network.

numpy to do the data mining, data cleaning and extract useful information. The data visualization is achieved by *seaborn* and *matplotlib*, and the random forest model is retrieved from *sciki-learn*.

This original dataset is repeated 20 times with random partitioning (i.e., 20 × cross-validation) to separate them into the training set (50%) as chosen, and testing set (50%) as the remaining based on Algorithm I. Then, based on Eqs. (5.2)–(5.3), 558 equivalence classes are derived out of the 7405 records, however, many of them only have one or two service records, which is not convincing for determining the probability (P). Therefore, the authors set a threshold value of 10 to filter the ones below it, and 273 equivalence classes out of 5655 records are selected based on Algorithm II. Then, by utilizing Eq. (5.5), $P(s)$ of each class can be calculated

New IT-driven value co-creation mechanism 111

to be compared with the C_L and C_U. The ones bigger than C_U are put into the accept region, in-between C_L and C_U in the pending region, and smaller than C_L in the reject region, respectively, based on Eqs. (5.19) −(5.21). Hence, according to Eq. (5.22) and Algorithm III, the average total cost (in red) and number of people rewarded (in blue) with different combination of C_L and C_U (interval of 0.05) is depicted in the line chart of Fig. 5.10A $C_L = 0.5$, and (b) $C_U = 0.5$, by calculating both the training (solid line) and testing dataset (dashed line), repeating 20 times. One should notice that the situations where $C_U = 0$, or $C_L = 1$ are not considered since they did not even exist according to Eqs. (5.17)−(5.18).

From the figure, one can find that the minimum total cost lies in where $(C_L, C_U) = (0.5, 0.6)$, while the maximum users get awarded lies in $(C_L, C_U) = (0.5, 0.5)$. The service providers can utilize this data to determine their marketing strategies with budget constraints and coverage concerns. For example, to balance the total cost with maximum people rewarded, so as to encourage active participations cost-effectively. Meanwhile, the line chart of training set (solid line) and testing set (solid line) in Fig. 5.10 have the similar curve, which help approve the accuracy of original random forest established for prediction purposes.

Moreover, according to three-way incentive model, another point is to discover whether the training dataset is well enough for future service innovation prediction based on the conditions given. In this experiment, 50%, 60%, 70%, 80%, and 90% of the original dataset was randomly chosen as the training set, with the remaining as the testing set. The testing result can be found in Table 5.4, where the accuracy is determined based on the value of precision (i.e., accuracy) and recall (i.e., sensitivity). One can find that the dataset is not that robust with a very high accuracy, which can result from the sample size of each equivalence class or the lack of specific

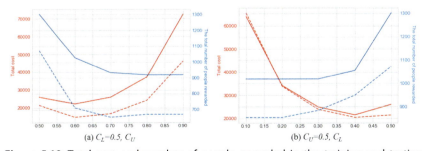

Figure 5.10 Total cost and number of people rewarded in the training and testing dataset.

Table 5.4 Prediction accuracy result.

Percentage of training dataset	Precision (%)	Recall (%)
50%	76	83
60%	81	85
70%	82	84
80%	95	95
90%	90	91

combination of condition attributes in the dataset. Due to confidentiality, the authors cannot access more data from the company. Nevertheless, the result still reflects the approach is good enough to deal with scalability.

5.4 Chapter summary

This chapter discussed the IT-driven technical issues to enable the value co-creation of Smart PSS development from both the product-service level and system level, respectively.

For the product-service level development, the relevant works on value co-creation design theory, IT-driven toolkits, and the concept of SCOAP were discussed in detail. Furthermore, the modeling technique of the SCP and its services were given lastly, which serve as the foundation for Smart PSS solution design with active user participation.

For the system level development, the state-of-the-art MCS approach was leveraged with hybrid intelligence to enable the industrial value co-creation in a cost-effective manner. The incentive mechanism, generic system framework, data collection and fusion manner, and the cost-driven decision making process for value generation were depicted, together with an illustrative example of the SWD maintenance service in the MCS network environment.

It is hoped this chapter can offer a systematic and promising manner for value co-creation and design innovation of Smart PSS by leveraging both human intelligence and machine intelligence, and can provide useful insights to service providers/manufacturers for their own value co-creation process implementation in a cost-effective manner.

References

Bénade, M., Brun, J., Le Masson, P., & Weil, B. (2016). How smart products with built in flexibility empower users to self - design the use: A theoretical framework for use generation. In *14th open and user innovation conference, Boston, United-States*. Retrieved from https://hal.archives-ouvertes.fr/hal-01389650.

Faltings, B., Li, J. J., & Jurca, R. (2014). Incentive mechanisms for community sensing. *IEEE Transactions on Computers, 63*(1), 115−128. https://doi.org/10.1109/TC.2013.150

Franke, N., & Piller, F. (2004). Value creation by toolkits for user innovation and design: The case of the watch market. *Journal of Product Innovation Management, 21*(6), 401−415. https://doi.org/10.1111/j.0737-6782.2004.00094.x

Ganti, R. K., Ye, F., & Lei, H. (2011). Mobile crowdsensing: Current state and future challenges. *IEEE Communication*, 32−39. https://doi.org/10.1109/MCOM.2011.6069707

Gross, U., & Antons, D. (2009). *Embedded open toolkits for user innovation: Postponing new product development decisions into the customer domain. Wi* (pp. 835−840).

Gu, P., Hashemian, M., & Nee, A. Y. C. (2004). Adaptable design. *CIRP Annals - Manufacturing Technology, 53*(2), 539−557. https://doi.org/10.1016/S0007-8506(07)60028-6

Guo, K., Tang, Y., & Zhang, P. (2017). CSF: Crowdsourcing semantic fusion for heterogeneous media big data in the internet of things. *Information Fusion, 37*, 77−85. https://doi.org/10.1016/j.inffus.2017.01.008

Guo, B., Wang, Z., Yu, Z., Wang, Y., Yen, N. Y., Huang, R., & Zhou, X. (2015). Mobile crowd sensing and computing. *ACM Computing Surveys, 48*(1), 1−31. https://doi.org/10.1145/2794400

Gu, P., Xue, D., & Nee, A. Y. C. (2009). Adaptable design: Concepts, methods, and applications. *Proceedings of the Institution of Mechanical Engineers - Part B: Journal of Engineering Manufacture, 223*(11), 1367−1387. https://doi.org/10.1243/09544054JEM1387

Hatchuel, A., & Weil, B. (2002). *C-K theory: Notions and applications of a unified design theory* (pp. 15−16). Retrieved from https://pdfs.semanticscholar.org/fa10/482827c6210 71896173dee9da0e8b427e39a.pdf.

Hatchuel, A., & Weil, B. (2009). C-K design theory: An advanced formulation. *Research in Engineering Design, 19*(4), 181−192. https://doi.org/10.1007/s00163-008-0043-4

Howe, J. (2006). The rise of crowdsourcing. *Wired Magazine, 14*(6), 1−5. https://doi.org/10.1086/599595

Hu, S. J. (2013). Evolving paradigms of manufacturing: From mass production to mass customization and personalization. *Procedia CIRP, 7*, 3−8. https://doi.org/10.1016/j.procir.2013.05.002

Jaimes, L. G., Vergara-Laurens, I. J., & Raij, A. (2015). A survey of incentive techniques for mobile crowd sensing. *IEEE Internet of Things Journal, 2*(5), 370−380. https://doi.org/10.1109/JIOT.2015.2409151

Jordan, K. O., Sheptykin, I., Grüter, B., & Vatterrott, H.-R. (2013). Identification of structural landmarks in a park using movement data collected in a location-based game. In *Proceedings of the first ACM SIGSPATIAL international workshop on computational models of place - COMP '13* (pp. 1−8). https://doi.org/10.1145/2534848.2534853

Kevin, A. (2009). That 'internet of things' thing. *RFID Journal, 22*(7), 97−114.

Koren, Y., Hu, S. J., Gu, P., & Shpitalni, M. (2013). Open-architecture products. *CIRP Annals - Manufacturing Technology, 62*(2), 719−729. https://doi.org/10.1016/j.cirp.2013.06.001

Koren, Y., Shpitalni, M., Gu, P., & Hu, S. J. (2015). Product design for mass-individualization. *Procedia CIRP, 36*, 64−71. https://doi.org/10.1016/j.procir.2015.03.050

114   Smart Product-Service Systems

Lane, N. D., Eisenman, S. B., Musolesi, M., Miluzzo, E., & Campbell, A. T. (2008). Urban sensing: Opportunistic or participatory?. In *HotMobile '08, proceedings of the 9th workshop on mobile computing systems and applications, February 26–26, 2008, Napa Valley, CA, USA* (pp. 11–16). https://doi.org/10.1145/1411759.1411763

Lee, J. S., & Hoh, B. (2010). Dynamic pricing incentive for participatory sensing. *Pervasive and Mobile Computing, 6*(6), 693–708. https://doi.org/10.1016/j.pmcj.2010.08.006

Lee, J., Kao, H. A., & Yang, S. (2014). Service innovation and smart analytics for industry 4.0 and big data environment. *Procedia CIRP, 16*, 3–8. https://doi.org/10.1016/j.procir.2014.02.001

Luo, T., & Tham, C.-K. (2012). *Fairness and social welfare in incentivizing participatory sensing.* Retrieved from http://arxiv.org/abs/1411.5795.

Ma, H., Zhao, D., & Yuan, P. (2015). Opportunities in mobile crowd sensing. *Infocommunications Journal, 7*(2), 32–38. https://doi.org/10.1109/MCOM.2014.6871666

Ota, K., Dong, M., Gui, J., & Liu, A. (2018). QUOIN: Incentive mechanisms for crowd sensing networks. *IEEE Network*, 114–119. https://doi.org/10.1109/MNET.2017.1500151

Parida, V., Sjödin, D., & Reim, W. (2019). Reviewing literature on digitalization, business model innovation, and sustainable industry: Past achievements and future promises. *Sustainability (Switzerland), 11*(2). https://doi.org/10.3390/su11020391

Pawlak, Z. (1982). Rough sets. *International Journal of Computer & Information Sciences, 11*(5), 341–356. https://doi.org/10.1007/BF01001956

Piller, F., Ihl, C., & Steiner, F. (2010). Embedded toolkits for user Co-design: A technology acceptance study of product adaptability in the usage stage. In *2010 43rd Hawaii international conference on system Sciences* (pp. 1–10). IEEE. https://doi.org/10.1109/HICSS.2010.178.

Piller, F. T., Ihl, C., & Vossen, A. (2010). A typology of customer Co-creation in the innovation process. *SSRN Electronic Journal.* https://doi.org/10.2139/ssrn.1732127

Piller, F. T., & Walcher, D. (2006). Toolkits for idea competitions: A novel method to integrate users in new product development. *R and D Management, 36*(3), 307–318. https://doi.org/10.1111/j.1467-9310.2006.00432.x

Porter, M. E., & Heppelmann, J. E. (2014). How smart, connected products are transforming competition. *Harvard Business Review, 92*(11), 64–88.

Porter, M. E., & Heppelmann, J. E. (2015). *Connected products are transforming companies how smart* (pp. 1–9). https://doi.org/10.1017/CBO9781107415324.004

Rijsdijk, S. A., & Hultink, E. J. (2009). How today's consumers perceive tomorrow's smart products. *Journal of Product Innovation Management, 26*(1), 24–42. https://doi.org/10.1111/j.1540-5885.2009.00332.x

Shi, W., Cao, J., Zhang, Q., Li, Y., & Xu, L. (2016). Edge computing: Vision and challenges. *IEEE Internet of Things Journal, 3*(5), 637–646. https://doi.org/10.1109/JIOT.2016.2579198

Simpson, T. W. (2004). Product platform design and customization: Status and promise. *Artificial Intelligence for Engineering Design, Analysis and Manufacturing, 18*(1), 3–20. https://doi.org/10.1017/S0890060404040028

Simpson, T. W., Jiao, J. R., Siddique, Z., & Hölttä-Otto, K. (2014). *Advances in product family and product platform design: Methods & applications.* https://doi.org/10.1007/978-1-4614-7937-6

Simpson, T. W., Maier, J. R., & Mistree, F. (2001). Product platform design: Method and application. *Research in Engineering Design, 13*(1), 2–22. https://doi.org/10.1007/s001630100002

Singer, D. J., Doerry, N., & Buckley, M. E. (n.d.). *What is set-based design?* Retrieved from http://www.doerry.org/norbert/papers/SBDFinal.pdf.

Sjödin, D., Parida, V., Kohtamäki, M., & Wincent, J. (2020). An agile co-creation process for digital servitization: A micro-service innovation approach. *Journal of Business Research, 112*(January), 478—491. https://doi.org/10.1016/j.jbusres.2020.01.009

Tortorella, G. L., Marodin, G. A., Fettermann, D. de C., & Fogliatto, F. S. (2016). Relationships between lean product development enablers and problems. *International Journal of Production Research, 54*(10), 2837—2855.

Tseng, M. M., Jiao, R. J., & Wang, C. (2010). Design for mass personalization. *CIRP Annals - Manufacturing Technology, 59*(1), 175—178. https://doi.org/10.1016/j.cirp.2010.03.097

Xue, D., Hua, G., Mehrad, V., & Gu, P. (2012). Optimal adaptable design for creating the changeable product based on changeable requirements considering the whole product life-cycle. *Journal of Manufacturing Systems, 31*(1), 59—68. https://doi.org/10.1016/J.JMSY.2011.04.003

Xu, Z., Mei, L., Choo, K. K. R., Lv, Z., Hu, C., Luo, X., & Liu, Y. (2017). Mobile crowd sensing of human-like intelligence using social sensors: A survey. *Neurocomputing, 279*, 3—10. https://doi.org/10.1016/j.neucom.2017.01.127

Yang, H., Deng, Y., Qiu, J., Li, M., Lai, M., & Dong, Z. Y. (2017). Electric vehicle route selection and charging navigation strategy based on crowd sensing. *IEEE Transactions on Industrial Informatics, 13*(5), 2214—2226. https://doi.org/10.1109/TII.2017.2682960

Yang, D., Xue, G., Fang, X., & Tang, J. (2012). Crowdsourcing to smartphones. In *Proceedings of the 18th annual international conference on mobile computing and networking - Mobicom '12* (p. 173). https://doi.org/10.1145/2348543.2348567

Zhang, B., Liu, C. H., Tang, J., Xu, Z., Ma, J., & Wang, W. (2018). Learning-based energy-efficient data collection by unmanned vehicles in smart cities. *IEEE Transactions on Industrial Informatics, 14*(4), 1. https://doi.org/10.1109/TII.2017.2783439

Zhang, H. R., & Min, F. (2016). Three-way recommender systems based on random forests. *Knowledge-Based Systems, 91*, 275—286. https://doi.org/10.1016/j.knosys.2015.06.019

Zheng, P., Lin, Y., Chen, C. H., & Xu, X. (2018). Smart, connected open architecture product: An IT-driven co-creation paradigm with lifecycle personalization concerns. *International Journal of Production Research.* https://doi.org/10.1080/00207543.2018.1530475

Zheng, P., Xu, X., & Chen, C.-H. (2018). A data-driven cyber-physical approach for personalised smart, connected product co-development in a cloud-based environment. *Journal of Intelligent Manufacturing, 1*—16. https://doi.org/10.1007/s10845-018-1430-y

Zheng, P., Wang, Z., Chen, C.-H., & Pheng Khoo, L. (2019). A survey of smart product-service systems: Key aspects, challenges and future perspectives. *Advanced Engineering Informatics, 42*(April), 100973. https://doi.org/10.1016/j.aei.2019.100973

Zheng, P., Xu, X., Yu, S., & Liu, C. (2017). Personalized product configuration framework in an adaptable open architecture product platform. *Journal of Manufacturing Systems, 43*, 422—435. https://doi.org/10.1016/j.jmsy.2017.03.010

CHAPTER 6

Graph-based context-aware product-service family configuration

Contents

6.1 Product-service family configuration in smart PSS	118
6.1.1 Context and context awareness in smart PSS	120
6.1.2 Context-aware solution configuration in smart PSS	121
6.2 Graph-based product-service-context modeling	123
6.2.1 Heterogeneous data collection and storage	125
6.2.1.1 The involved data in smart PSS	*125*
6.2.1.2 Data storage in smart PSS	*125*
6.2.2 Construct ontologies to organize the concepts	126
6.2.3 Graph model construction	127
6.2.3.1 Requirement graph construction	*128*
6.2.3.2 Smart PSS configuration hypergraph construction	*130*
6.2.4 Application layer	132
6.3 Requirement management based on the graph model	132
6.3.1 Requirement graph	133
6.3.2 Sequence order via random walk	134
6.3.3 Node embeddings via SkipGram	135
6.3.4 Predict the most relevant nodes	135
6.4 Solution configuration based on the hypergraph model	137
6.4.1 Hypergraph model	137
6.4.2 Offline training: an unbiased hypergraph ranking approach	138
6.4.3 Online application: user queries identification	140
6.5 Case study	140
6.5.1 Case study of the requirement management based on requirement graph	141
6.5.1.1 Requirement graph construction	*141*
6.5.1.2 Requirement elicitation	*143*
6.5.1.3 Results discussion	*144*
6.5.2 Case study of solution configuration based on the hypergraph	145
6.5.2.1 Hypergraph construction	*145*
6.5.2.2 Solution configuration	*146*
6.5.2.3 Results discussion	*147*
6.6 Chapter summary	147
References	148

Smart Product-Service Systems
ISBN 978-0-323-85247-0
https://doi.org/10.1016/B978-0-323-85247-0.00005-0

© 2021 Elsevier Inc.
All rights reserved.

In today's competitive market, accurately extracting users' requirements and further recommending the proper solutions will largely determine the final success of a Smart PSS. This process, as named Smart PSS configuration, consists of two key activities: requirement management and solution recommendation.

Requirement management intends to detect the end-user's requirements in an IoT-enabled environment (e.g., a cloud-based crowdsourcing platform). In this environment, a great number of user-generated data are accessible, such as user reviews and comments. The data contains valuable user experience information, and hence can be used for the requirement identification in a data-driven manner, and eventually serves as the guiding star for marketing people and design engineers to understand the trend in the market.

Simultaneously, since users would like to receive personalized solutions rather than unified ones in Smart PSS, solution recommendation is expected to select the most appropriate solution according to different user's request. Unlike the conventional configuration systems which largely rely on the user's professional knowledge, in Smart PSS configuration system, a feasible solution can be generated based on the usage scenario.

Despite the advantages of these two key activities, how to systematically organize the massive data and information from different resources remain as a major challenge. Therefore, it is necessary to establish a context-aware system framework for Smart PSS configurations, and the development teams should also have the following concerns:

- What information is involved in the Smart PSS configuration activities?
- How to organize the relevant information from the fundamental resource layer to the higher application layer?
- Which kind of model is suitable to handle the multipartite information in Smart PSS configuration?
- How to realize the context awareness feature in Smart PSS configuration process?

This chapter aims to provide a graph-based context-aware Smart PSS configuration framework, together with the proper approaches for its requirement elicitation and solution recommendation, which are elaborated below.

6.1 Product-service family configuration in smart PSS

Different from the other existing configuration systems, the Smart PSS configuration system should have the following three key features.

Firstly, the changing/upgrade of the embedded software, as a service, often replaces the modification of physical components. This improvement makes the configuration process in Smart PSS more flexible with various service options, which is reflected in its heterogeneity and dynamics. Specifically, the heterogeneity refers to the configuration capability of dealing with product components, service modules, and other information simultaneously. The dynamics refers to the iterative configuration capability that the PSBs nowadays can be agilely upgraded and ever-evolving instead of just configuring the PSBs once for all and finally recycling the total product at its end-of-life. As a result, more attention will be paid to the usage phase for iterative configuration to maintain competitiveness.

Secondly, the latent user requirement identification manner based on user-generated data replaces the explicit requirement identification manner dependent on the engineers' empirical knowledge. Contrary to the explicit requirements, the latent requirements refer to the requirements not directly expressed by the users. This type of requirement often exists in user-generated content. For example, a user comment of a 3D printer down-loaded from an e-commerce platform is "I am noticing that the bed is higher in the middle than it is as the four corners .... I finally got it all operational with a new hot end." In this case, the explicit requirement can be "changing a hot end". Although the user solved it by changing a new hot end, his latent requirement is "print bed is not flat" and can be solved by swapping the tool head for a Titan unit. This phenomenon happens because of the lack of domain knowledge, which can be solved by mining the semantic relations among the domain concepts. It is also one of the critical steps of latent requirement identification. The transformation into latent requirement extraction (1) reduces the subjectivity of users; (2) enriches the latent relations between the domain concepts; and (3) releases the stress of engineers to correctly extract requirements from a group of the target users.

Thirdly, dynamically mapping the personalized solution recommendations in a context-aware manner replaces the conventional solution recommendation approach in the function-oriented manner. The functions usually significantly differentiate the solutions, especially for the solutions in the industry. However, the inaccurately selected functions due to the lack of domain knowledge will always lead to mismatched solutions, thus reducing the users' satisfaction and experience. To alleviate this problem, some auxiliary information should be involved in the Smart PSS config-uration process. Usage scenario information, as one of the accessible information in Smart PSS, is the information users will always correctly know while submitting the user queries. Hence, by adding the usage scenario as the auxiliary information, Smart PSS configuration will become more user friendly with reliable user inputs and personalized results.

As a summary, the Smart PSS configuration is expected to be (1) ever-evolving in the usage phase; (2) able to extract latent requirements; and (3) capable to select personalized solutions with usage scenarios as the auxiliary information.

To achieve these goals, a context-aware configuration system is required. In recent years, numerous applications under the IoT have been developed based on context awareness, such as personalized recommendation (Champiri, Shahamiri, & Salim, 2015), smart home (Uddin, Khaksar, & Torresen, 2018), and smartphone services (Chen, Chu, Su, & Chu, 2014). However, the implications of context-aware systems in PSS has not yet been studied, let alone the context-aware Smart PSS configuration system.

6.1.1 Context and context awareness in smart PSS

To detect the usage scenarios and react to their changes, a context-aware system is expected in the Smart PSS configuration. It is originally a concept from computer science, which was first proposed by Schilit, Adams, and Want (1994). They specifically defined context awareness as the capability to detect and react to the information about users and the related IT devices, such as the environment, situation, location, and other surrounding information (Schmidt et al., 1999), whereas other researchers also gave broader or more specific definitions of context awareness. According to the foundational definition from Dey (2001), a system is context-aware if it uses context information to offer products or services to the users, in which the dependency of context information depends on the users' tasks. Kim et al. (2004) broadly defined context awareness as the capability of acquiring and applying context. Despite the different definitions, the purpose of the context awareness is still consistent: to offer users/devices/environment's information to the computer, and then to adjust the system's reactions.

Since a context-aware system needs to percept the context, it is necessary to clarify what the context is and what information belongs to context. There are many taxonomies to the context. Schmidt et al. (1999) divided the context into three classifications: (1) *computing context*, such as the communication bandwidth, network connections; (2) *user context*, such as the user's profile, user preference, and historical behavior; and (3) *physical context*, such as lighting, noise, and temperature. Chen and Kotz (2000) described the context as a set of environmental factors and their states that can identify either the user's task or in which situation the task happen. Nowadays, a widely referred definition of context was proposed by Abowd et al. (1999):

> *Context is any information that can be used to characterize the situation of an entity. An entity is a person, place, or object that is considered relevant to the interaction between a user and an application, including the user and applications themselves.*

Through the discussion, we can find context is a broad concept. To the authors' knowledge, the context can be grouped as explicit context and the implied context. The explicit context contains the physical information, such as location, time, device states; meanwhile, the implied context denotes the user profile, behavior, preferences, and other user-related information.

6.1.2 Context-aware solution configuration in smart PSS

To establish context-aware applications, the involved entities and their relationships should be predefined (Teixeira, Maran, de Oliveira, Winter, & Machado, 2019). A well-defined context model forms the basis of context-aware applications (Stokic & Correia, 2015). The context information in Smart PSS can be grouped from the perspective of data sources and data format.

- From the perspective of data sources, the context data can be collected from users and sensors.

  The context data from users includes user requirements, user comments, the figures/audio/video generated by users. The context data from sensors are broad and diverse, including the location, weather, user behavior data (e.g., posture parameters if wearable devices are involved) and machine status data.

- From the data format perspective, the context data are mainly separated as structured data (e.g., numerical data), semistructural data (e.g., table) and unstructured data (e.g., natural language).

  Numerical context data are required several methods to deal with as well. Some numerical context features can be identified and grouped based on common knowledge/domain knowledge. For instance, time as a physical context can be easily predefined as {morning : 7 : 01 −12 : 00, noon : 12 : 01 −13 : 59, afternoon : 14 : 00 −17 : 59, ...}.

As for the semistructural data which embraces context data, we can predefine the contexts based on table headers and table elements/annotations. For example, suppose we have a user-generated table containing the username, age, and occupation, then we can obtain the user contexts through the table headers and the values inside the tables. Besides, the phrases and sentences generated by users also embrace much context information, they are the unstructured context data.

It is intuitive to handle those data with context-aware computing techniques, whose basic processes can be separated into three phases, namely context acquisition, context processing, and context usage (Barrett & Power, 2003). Context acquisition, as the first phase, identifies context information via sensors or user interfaces. Physical contexts can be collected via sensors, such as the location information from GPS, the time information from the computers' built-in clock, and the luminance information from the photosensitive diode. Whereas advanced and implied context can be collected via user interface. Then context processing will transform the context information into the useful usage scenario patterns. Finally, the system will apply the context information to adjust itself or give relevant responses.

In Smart PSS, industrial solutions' configuration is expected to be sufficiently flexible to meet various usage scenarios. The Smart PSS configuration can refer to selecting proper solutions based on usage scenario information to meet the user requirements. The configuration process involves a fundamental modeling task and two critical design tasks:

(1) Modeling the customer requirements, the PSBs information, and their relationships into a compatible manner;
(2) Identification of key information of customer requirements; and
(3) The personalized mapping between customer requirements and product-service bundles (PSBs) in terms of various usage scenarios.

To clarify the research scope, some assumptions of Smart PSS configuration are predefined as follows:

- It is assumed that the initial product-service family has already been launched to the market so that the historical configuration orders or user reviews are accessible.
- It is assumed that while configuring, all the usage scenarios can be grouped into a current usage scenario group. The situation that a new usage scenario appears is out of the research scope.

- It is assumed that the PSBs have been modularized before launching to the market; meanwhile the modularization results of PSBs will not change during the usage phase. Thus, the modularization process is out of the research scope.

6.2 Graph-based product-service-context modeling

In order to organize the diverse information from different data sources, a fundamental model should be constructed, which can answer the first three research questions proposed at the beginning of this chapter. The authors selected product components information, service modules information, and the context information to build up the graph-based model, owing to their significance in Smart PSS.

Graph-based models have already been widely used in many applications of the product and/or service systems (Järvenpää, Siltala, Hylli, & Lanz, 2019; Li, Tao, Cheng, Zhang, & Nee, 2017), for example, the representation of the complex system (Wu et al., 2018), PSS modularization (Li, Chu, Chu, & Liu, 2014), and PSS solution recommendation (Huang, Fan, & Tan, 2014). Kim, Wang, Lee, and Cho (2009) used graphs to represent PSS, consisting of values, products, services, and their relations. The graph-based PSS representation was proved feasible via an example of a meal assembly kitchen under complex usage scenarios. At the same time, the graph was also applied in design decision tasks. Wang, Chen, Zheng, Li, and Khoo (2019) adopted a graph-based approach for requirement elicitation to release the engineers' stress on manually extracting requirements based on their empirical knowledge. Zheng, Chen, and Shang (2019) also developed a DSM method to automatically manage the engineering changes in Smart PSS. All the examples validate the strength of graph to represent heterogeneous entities and complex relationships in Smart PSS.

Hence, a graph-based Smart PSS configuration framework is proposed in this section, as shown in Fig. 6.1. It is a five-layered structure, including *resource layer, knowledge management layer, graph construction layer*, and *application layer*.

Resource layer contains SCP, its generated services, sensor networks, and databases, serving as the physical resources. Simultaneously, product attribute data, service logs, and historical requirement specification documents constitute the data resources, corresponding to the physical resources.

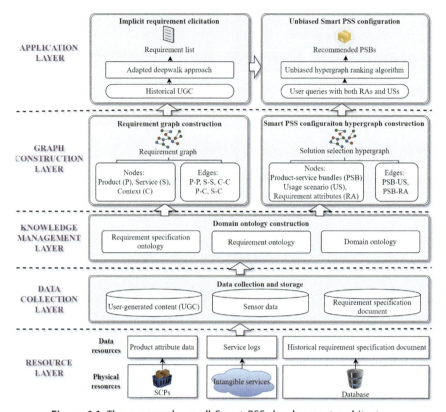

Figure 6.1 The proposed overall Smart PSS development architecture.

Data collection layer collects and stores the heterogeneous data from various data sources. Both the *resources layer* and *data collection* layer serve as the basis of the Smart PSS configuration framework by upholding the other upper layers with multi-source data.

Knowledge management intends to systematically organize the information. This layer defines three ontologies, namely requirement specification ontology, requirement ontology, and domain ontology. Note that the professional concepts in domain ontology are case-by-case; hence an approach to extracting the critical domain concepts and organizing them hierarchically is necessary.

Application layer depicts the process of eliciting requirements based on the requirement graph, selecting proper PSBs based on the Smart PSS configuration graph, and finally checking the status of the usage scenarios for further design iteration or operation.

6.2.1 Heterogeneous data collection and storage

6.2.1.1 The involved data in smart PSS

Fig. 6.1 shows that the data collection layer contains user-generated data content (UGC), sensor data, and requirement specification document. Those data come from different sources, which are users, sensors, and documents, respectively.

- UGC refers to the data created by users via online IT platforms, including numbers (e.g., user ratings), structured text (e.g., tables), unstructured text (e.g., product comments and Q&A discussion), and images. It embraces lots of meaningful concepts with semantics, thus making itself a powerful data source for requirement extraction and solution configuration. Plenty of critical concepts about product and service performance, such as good quality, lightweight, fast logistic, and excellent customer service, can be extracted, which is useful for context acquisition. By extracting those domain concepts and their semantic relations, the UGC can be regarded as the social sensors since it represents human intelligence.
- Sensor data refers to the signals collected from SCPs. Some typical sensor data contains acceleration, angular velocity, location, and illumination intensity. The authors use them for the context acquisition as well.
- Requirement specification document denotes the historical requirement lists from the service providers. It is collected as a requirement reference for the later modules.

6.2.1.2 Data storage in smart PSS

To efficiently represent the entities' properties and relations, NoSQL database is further recommended in this framework. Specifically, the extracted domain concepts will be stored in NoSQL databased with a graph-based format whose nodes stand for the involved instances in the system and edges stand for their relationships. The authors recommend the graph-based format because of its innate strength in representing the manifold many-to-many relations. Compared with the relational database, the graph database does not need to create a datasheet while adding new types of relations, thus making it more efficient to represent the manifold relations. Meanwhile, graph database requires no strict restrictions on developers to well-organize the data relations, thus outperforming on the many-to-many relations. Nowadays, a typical graph database, Neo4j, has good performance for retrieval and semantization, which is a good candidate for data storage in Smart PSS configuration.

6.2.2 Construct ontologies to organize the concepts

To organize domain concepts as reference systematically, the knowledge management layer containing ontologies is established.

Both the requirement identification and solution recommendation in Smart PSS are knowledge-intensive, engineers would like to organize the knowledge patterns for knowledge reuse. Owing to the strength in knowledge structure representation from upper-level knowledge to domain knowledge, ontology is used as a knowledge base in this layer.

Three ontologies are defined in Smart PSS configuration: requirement specification document ontology, requirements ontology, and application domain ontology (Amaxilatis et al., 2018). Fig. 6.2 shows their relationship. Specifically, requirement specification document ontology organizes the structure of the requirement specification documents. Requirement ontology offers the requirement taxonomy, such as functional requirements and nonfunctional requirements. Meanwhile, domain ontology addresses the relations among the domain concepts.

When constructing the domain ontology, automatic methods are much more efficient, but it is hard to be achieved if no ready-made domain ontology is given. Hence, a semiautomatic method is adopted in this framework. The raw domain knowledge can be extracted from

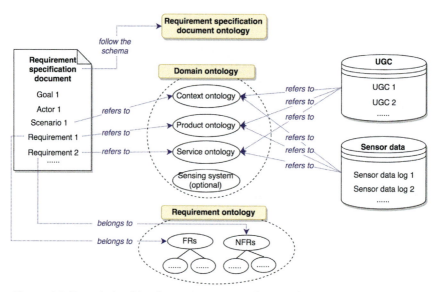

Figure 6.2 The relationship of requirement specification document and ontologies.

requirement specification documents, product configuration documents, and user-generated data. The product-related concepts are usually about material, product parameter, functions and so on. Associate services can be predicted maintenance, personalization, postprocessing, and so on. All the concepts and their upper-lower relationships, including Hypernym (is a kind of ...), Hyponym (... is a kind of), Holonym (is a part of ...), Meronym (... is a part of) can be identified via text mining techniques.

6.2.3 Graph model construction

Recap that we have already discussed graph models' applications in product and/or service systems at the beginning of Section 6.2, frequently mentioned examples include complex system representation, modularization, and reasoning tasks. This subsection will expound on why the graph model is selected and how the related information establishes the graphs in Smart PSS.

Numerous heterogeneous entities constitute the Smart PSS; meanwhile, the multiple data sources differ the data formats, thus there is a necessity to integrate the multipartite entities and their relations into a uniform model. Graph models can intuitively represent multipartite entities and their relations. Particularly, the data itself on a graph can show their relationship, unlike clarifying their relationship through the codes conventionally. Hence graph model facilitates the transformation from "smart application code" into "smart data." Based on this strength, the graph model is selected to formalize the Smart PSS design activities in this study (Fig. 6.3).

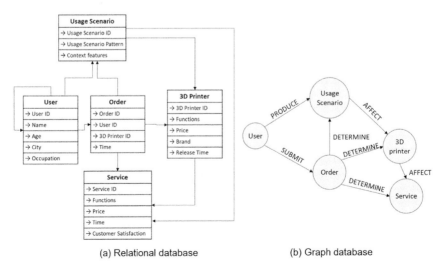

Figure 6.3 The comparison between relational data model and graph model (Team., n.d.).

6.2.3.1 Requirement graph construction

As shown in Fig. 6.1, firstly, a requirement graph (RG) denoted as $RG = \langle V, E \rangle$ is established for the requirement elicitation. It is composed of three types of nodes, namely product components (P), service modules (S), and context (C). Hence, the vertex set $V = P \cup S \cup C$ is the union of three types of nodes. Meanwhile, edge set E is the union of five subsets CP, CS, PP, SS and CC. All the nodes are extracted from UGC, sensor data, and historical requirement specification documents, which are preprocessed in the domain ontologies. Their meanings are explained as follows.

- **Products components (P)** means the physical products' components or modules, such as motor, seat, nozzle, and so on. They are organized in a product ontology which looks like a tree diagram, where the component p_i refers to the ith components or modules.
- **Services modules (S)** refers to the services offered by the company, such as predicted maintenance, customer relation management, customized data analysis, and so on. Similar to the product components, they are organized in a service ontology, where the component s_i refers to the ith service modules. The services can be grouped into either *digitalized services* or *e-services*. *Digitalized services* refer to the integrated services which have intimate interactions with the physical products, such as the digital twin of a machine tool; while *e-services* are the software-based services independent with the physical products, for instance, the APPs in the smartphones (Zheng et al., 2019).
- **Context (C)**, as defined in Section 6.1.1, refers to the features to characterize the usage scenarios and their feelings during usage. For example, wind speed is an environmental factor affecting the riding experience; while the appearance and the installation instruction of a bike are two implied context which represents the users' feelings. A usage context is expressed as c_i, indicating the i-th context in the context ontology.

Besides the nodes, the edges in the RG should be defined as well. Their schema is shown in Fig. 6.4. If two concepts co-occur in the UGC/requirement specification documents or are predefined by engineers for sensor data collection, an edge should be created. Their meanings are explained in the following.

- **Product–Product relationship (PP)** refers to the cooccurrence between two product components in the data resources. This type of edges will commonly appear since many product components have correlations with each other.

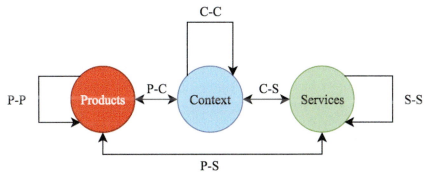

Figure 6.4 Schema of five different relationships between nodes.

- **Service-Service relationship (SS)** refers to the cooccurrence between two service modules in the data resources. SS relation will happen if they are similar or they are in sequential order.
- **Context-Context relationship (CC)** refers to the cooccurrence between two contexts. In practice, many contexts will be simultaneously mentioned in the product comments, it is necessary to imitate this relation as CC relation.
- **Product-Context relationship (PC)** indicates the cooccurrence between a product component and a context. This relationship will exist in the user comments when the users describe his/her usage scenarios.
- **Context-Service relationship (CS)** indicates the cooccurrence between a service module and a context.

So far, we have defined the elements in the RG; next we will define some matrices for mathematical representation and for the further reasoning steps.

A graph $G = (V, E)$ is essentially consisting of two sets, namely $V = \{v_1, v_2, v_3, \ldots\}$ as a set of vertices and $E = \{e_1, e_2, e_3, \ldots\}$ as a set of edges. An adjacent matrix A is defined as Eq. (6.1), whose elements denote whether node v_i and v_j are connected.

$$A_{ij} = \begin{cases} 1 & \text{if } v_i \text{ and } v_j \text{ are connected} \\ 0 & \text{otherwise} \end{cases} \quad (6.1)$$

Furthermore, a diagonal matrix D_v can also be identified, whose diagonal elements denote the degree of vertex k_i that $k_i = \sum_{j=1}^{n} A_{ij}$.

6.2.3.2 *Smart PSS configuration hypergraph construction*

A hypergraph is a generalization of a graph that uses hyperedges to connect a set of nodes rather than constant two nodes (Zhou, Huang, & Schölkopf, 2007). Despite the plentiful examples of graph-based models in society and nature, the study of using hypergraph to formalize the Smart PSS configuration tasks is still rare. Hypergraph fits for the solution configuration owing to its strength in representing multipartite entities and the many-to-many relations without information loss.

Conventionally, a development team relies on functional requirement attributes (RAs) to select a proper PSB as a solution. However, to alleviate the lack of domain knowledge and inaccurate configuration queries, the usage scenario (US) is introduced into the Smart PSS configuration process. Both the functional RAs and the USs can be easily accessed in Smart PSS. The functional RAs come from the product specification documents, such as the requirements on material types, product size, brands, etc. The US information can be extracted from the key phrases in the product comments, such as fast delivery, nondestructive packaging, excellent quality and so on. Besides RAs and USs, the PSBs refers to the smart, connected products together with the services offered by the service providers. As a result, three types of information are used for the Smart PSS configuration. Mathematically, those entities are regarded as nodes in the hypergraph, denoted as $PSB = \{psb_1, psb_2, ...\}$, $US = \{us_1, us_2, ...\}$, and $RA = \{ra_1, ra_2, ...\}$.

Meanwhile, two hyperedges, namely $E^{(1)}$: Hyperedge PSB-RA and $E^{(2)}$: Hyperedge PSB-US, can be intuitively defined as well.

- $E^{(1)}$: **Hyperedge PSB-RA** means the relationship between product-service bundles and requirements attributes. This relationship usually appears in the product specification documents since PSBs are configured via different values on requirement attributes. At the same time, this relationship is usually composed of a PSB and several RAs because PSBs are nowadays always multifunctional.

- $E^{(2)}$: **Hyperedge PSB-US** indicates the relationship between product-service bundles and usage scenarios. It can only be collected from product comments after purchasing a PSB, so that the US can be generated and feed back in the comments. Similar to $E^{(1)}$, $E^{(2)}$ usually link a PSB with several USs as well since a PSB can be used in a variety of usage events/interactions/experiences.

In this way, a hypergraph $G = (V, E)$ can be established, whose schema is shown in Fig. 6.5.

To mathematically represent the hypergraph, some matrices should be clarified as well.

A hyperedge is incident with a vertex v when $v \in e$. It is defined that $h(v, e) = 1$ if a vertex v is in a hyperedge e, otherwise it equals 0, denoted as:

$$h(v_i, e_j) = \begin{cases} 1 & \text{if } v_i \in e_j \\ 0 & \text{otherwise} \end{cases} \tag{6.2}$$

We can get the incident matrix $H \in \mathbb{R}^{|V| \times |E|}$ whose elements are $h(v_i, e_j)$. The degree of a vertex v is $d(v) = \sum_{v \in e} w(e) h(v, e)$, referring to how many hyperedges are connected by the vertex. Unlike the ordinary graph, the degree of a hyperedge e is defined as $\delta(e) = \sum_{v \in V} h(v, e)$, indicating the node count in a hyperedge. Let $D_v \in \mathbb{R}^{|V| \times |E|}$ indicate the diagonal matrix, whose diagonal elements are the degrees of all nodes. Similarly, $D_e \in \mathbb{R}^{|E| \times |E|}$ is the diagonal matrix containing the hyperedge degrees. Meanwhile, we set $W \in \mathbb{R}^{|E| \times |E|}$ as the diagonal matrix whose diagonal elements are the hyperedges' weights.

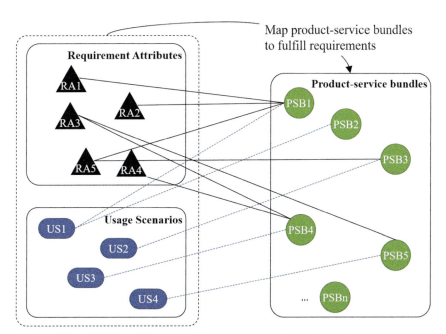

Figure 6.5 The objects and relations in the smart PSS configuration.

6.2.4 Application layer

Many design activities can be conducted based on the graph-based model, such as modularization, solution recommendation and so on. Here, two core design activities are studied, i.e., requirement elicitation and Smart PSS configuration. The relative reasoning algorithms are also developed to make wise decisions in Smart PSS.

In Smart PSS, both the explicit requirements and latent requirements are supposed to be explored. Unlike the conventional requirement extraction methods which only focus on the explicit expressions from users in a labor-intensive manner, the latent relations with semantic meanings are solicited in a data-driven manner by integrating usage contexts via a deepwalk-based approach. The construction of the RG is introduced in Section 6.2.3. The details of the requirement elicitation algorithm are explained in Section 6.3.

The extracted requirement lists can be used to enrich the hypergraph in the process of Smart PSS configuration. Meanwhile, the usage scenario information is used to enrich user queries and reduce the inputs' uncertainty. The Smart PSS configuration hypergraph construction is introduced in Section 6.2.3. The proposed unbiased hypergraph ranking algorithm is introduced in Section 6.4. The configuration logs containing users' final choices on the recommended PSBs enrich the RG and the hypergraph, making Smart PSS design in a closed loop.

6.3 Requirement management based on the graph model

This subsection will introduce how to extract requirements, especially latent requirements, based on the RG constructed in Section 6.3.

A requirement can be templated as "Under which contexts, system components(s) shall/should/will do process" according to Rupp's requirement template, as shown in Fig. 6.6. *Context* follows the aforementioned definition. *System components* represent product components or service components in Smart PSS. *Process* is the actions a product/service

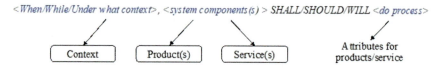

Figure 6.6 Requirement representation and its relations with smart PSS components.

can perform. As an illustration, a requirement can be written as "When CPU's temperature is high, the computer radiator should work," in which "CPU's temperature is high" serves as the context and "computer radiator" is the product components taking actions.

Like this, a requirement can be converted into the relations between context, product and service in Smart PSS. Accordingly, the requirement elicitation task can be transformed as mining the linkages between context(s), product(s), and service(s). To achieve it, a deepwalk-based approach is introduced in this subsection. Fig. 6.7 shows the overall flowchart of the latent requirement elicitation.

6.3.1 Requirement graph

The RG $RG = \langle V, E \rangle$ is the input of the deepwalk-based approach. Initially, product (P), service (S) and context (C) can be represented as one-hot encoding vectors using binary $\{0,1\}$ code. However, the one-hot encoding has two limitations. First, the vector dimensions can be too great if a huge amount of entities are involved in the RG. Specifically, if 10,000 entities are involved in the RG, then the vector will have 10,000 dimensions. Meanwhile, only one dimension is 1 in the vector but all the rest of the dimensions are 0, so the vertices' matrix $V \in \mathbb{R}^{|V| \times |V|}$ is extremely sparse and wastes lots of storage space. Second, one-hot encoding

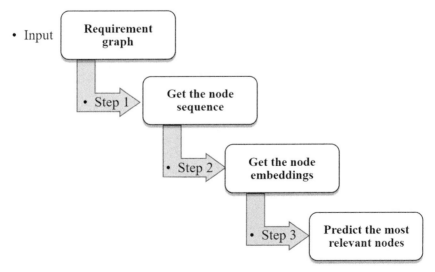

Figure 6.7 The overall flowchart of the proposed requirement elicitation approach.

loses the concepts' semantic relations in RG since the distances between any two nodes are always same.

To tackle these limitations, the node embedding technique is used in this approach for computation efficiency. We intend to represent nodes as low-dimensional vectors and hope that similar nodes will be close in the high dimensional space. This type of node embeddings is much more efficient for the machine learning tasks, for example, linkage prediction. The steps of reducing the node vectors' dimensions and predicting linkages will be introduced in the following subsections.

6.3.2 Sequence order via random walk

Deepwalk is used to obtain the node embeddings, which basically has two components, namely a *random walk generator* and *SkipGram*.

As the first component, the random walk generator creates node sequences via truncated random walk, and it treats the node sequences as vectors which is rooted at vertex v_i within a window w, denoted as $W_{v_i} = \{v_{i-w}, v_{i-w+1}, \ldots, v_{i-w} -1, v_{i+w}, \}$. The transition process among nodes can be regarded as a stochastic process. Technically, the pseudocode of the random walk generator for the RG is demonstrated in Table 6.1.

Table 6.1 The pseudocode of a random walk generation procedure.

The algorithm architecture of a random walk generation procedure	
Input: walk_length l, start_node v_i, requirement graph RG	
Output: a Random walk W	
Initialisation:	
1:	Given adjacency matrix A to represent requirement graph RG
2:	Given walk_length l
3:	Given start node v_i
4:	Set an empty list W as a random walk
5:	Set l' as the current walk length in the random walk W
	Generate the random walk:
6:	WHILE $l' < l$ do:
7:	IF no. of current_node_neighbor >0 do:
8:	Append a random node among current_node_neighbor to the walk
9:	ELSE break
10:	END IF
11:	END WHILE
	RETURN W
12:	END

6.3.3 Node embeddings via SkipGram

As the second component of the Deepwalk algorithm, SkipGram was originally a language model to learn the word embeddings. Here we use it to learn the node embeddings based on the node sequences from the former step. A mapping function $\Phi : v \in V \mapsto \mathbb{R}^{|V| \times d}$ is introduced, where $d \ll |V|$. It aims to maximize the cooccurrence probability among the nodes in the window w, viz. $Pr(\{v_{i-w}, ..., v_{i-1}, v_{i+1}, ..., v_{i+w}|v_i\})$, referring to predict both the former and later nodes based on node v_i. In this way, the requirement elicitation task can be formulated as the prediction based on the relevant nodes of a given context, its objective function is defined as Eq. (6.3).

$$\min_{\phi} - \log Pr(\{v_{i-w}, ..., v_{i-1}, v_{i+1}, ..., w_{i-w}|v_i\}) \qquad (6.3)$$

Considering the node embeddings, the objective function is denoted as Eq. (6.4).

$$\min_{\phi} - \log Pr(\{\phi(v_{i-w}), ..., \phi(v_{i-1}), \phi(v_{i+1}), ..., \phi(v_{i-w})|\phi(v_i)\})$$

$$(6.4)$$

in which $\Phi(v_i) \in \mathbb{R}^d$ is the vertex v_i's embedding.

SkipGram model approximates the cooccurrence probability with an independence assumption, hence the cooccurrence probability can be written as Eq. (6.5).

$$Pr(\{v_{i-w}, ..., v_{i-1}, v_{i+1}, ..., w_{i-w}|\phi(v_i)\}) = \prod_{j=i-w}^{i+w} Pr(v_j|\phi(v_i)) \quad (6.5)$$

To improve computational efficiency, one will respectively use hierarchical Softmax and stochastic gradient descent (SGD) to approximate the probability distribution and to optimize the parameters.

6.3.4 Predict the most relevant nodes

After obtaining the node embeddings $\Phi(v_i)$, one should predict the linkages among the nodes. Technically, cosine similarity between nodes is adopted in the approach for the sake of algorithm complexity, as shown in Eq. (6.6).

$$S(v_i, v_j) = \cos(v_i, v_j) = \frac{\phi(v_i) \cdot \phi(v_j)}{|\phi(v_i)| \cdot |\phi(v_i)|} \qquad (6.6)$$

136 Smart Product-Service Systems

Table 6.2 shows the pseudocodes of the link prediction based on the cosine similarity. The linkages with top k similarities can be extracted as the results. These linkages embrace the semantic information from human intelligence, including the explicit requirements and latent requirements.

By leveraging the Deepwalk–based approach, when a specific node in the RG is queried, the system will trigger its related nodes which have high cooccurrence probability. Either direct or indirect edges connect these derived nodes and the original given node. If they are originally indirectly connected, then their linkage constitutes a latent requirement.

Table 6.2 The pseudocode of extracting similarities between nodes.

The algorithm architecture for extracting similarities between nodes

Input: Id of the start node, adjacency matrix A of the RG, vertex set V, and the embeddings set Φ of nodes
Output: Rank of similarity between nodes

Initialisation:	
1:	Given adjacency matrix A to represent requirement graph RG
2:	Given the id of the start node v_0, set the id as id_0
3:	Given the embeddings set Φ of nodes
4:	Set an empty list S as the rank of similarity between nodes

Compute the similarity between nodes:	
5:	Obtain the embedding Φ_o of the start node
6:	**For** node v_i in vertex set V **do**:
7:	Obtain the embedding Φ_i of the product node v_i
8:	Compute the similarity between the start context node and the current product node
9:	$s_{t_{ij}} = \dfrac{\Phi_o \cdot \Phi_i}{\lVert \Phi_o \rVert \cdot \lVert \Phi_i \rVert}$
10:	Append the current $s_{t_{ij}}$ to the list S
11:	**End for**

Rank the similarity score:	
12:	Rank the elements in list S from high to low
13;	**RETURN** rank of similarity between nodes
	END

6.4 Solution configuration based on the hypergraph model

Except for the requirement extraction, another critical activity in Smart PSS configuration is to select proper PSBs according to user queries for personalization, which is called solution configuration (Long, Wang, Shen, Wu, & Jiang, 2013). This subsection will introduce an unbiased hypergraph ranking algorithm based on the established hypergraph model. Also, the reasons for choosing hypergraph model for Smart PSS configuration will be expounded.

The conventional PSS configuration usually asks the users to identify their expected functions. However, due to the lack of professional domain knowledge on the complicated products, inaccurate function attributes will be offered by the users, thus leading to the unmatched PSBs recommended to them. This limitation reduces the accuracy of the PSS configuration and tarnishes the user experience.

Whereas, Smart PSS configuration aims to offer a more flexible configuration process by allowing users to express their expected usage scenarios using natural language. Although the users might lack professional function knowledge, they always know their expected usage scenarios before purchasing a PSB. By adding the usage scenario into user queries, the Smart PSS configuration process can reduce the user queries' inaccuracy, hence (1) increasing the performance on personalization and (2) improving the user experience.

Basically, the Smart PSS configuration approach has three phases: hypergraph model construction, offline training, and online application. Firstly, the design team needs to construct the hypergraph based on the historical user reviews, PSB specification documents, and the PSB list. Then offline training aims to derive an optimal ranking score function from selecting proper PSBs based on the given requirements. Finally, the online application clarifies how to identify the query vectors according to user requirements while using the ranking score function. The details of each phrase are expounded as follows.

6.4.1 Hypergraph model

Recall that the hypergraph model construction was introduced in Section 6.2.3.2. It contains three types of nodes, i.e., PSB, RA, and US, and two types of hyperedges, namely, PSB-RA and PSB-US. It serves as the foundation of the Smart PSS solution configuration; both the offline training process and the online application process perform based on it.

6.4.2 Offline training: an unbiased hypergraph ranking approach

To react to different user queries, the offline training part intends to learn a ranking score function $\mathbf{f} : V \to \mathbb{R}$ based on the hypergraph $G = (V, E)$ and the query vector $\mathbf{y} = \left[y_1, y_2, \ldots, y_{|V|}\right]^T$, where y_i is the initial scores of the query vectors (Tan et al., 2011). A cost function $Q(\mathbf{f})$ is defined to ensure that the ranking score function has the lowest loss, as shown in Eq. (6.7).

$$Q(\mathbf{f}) = \frac{1}{2} \sum_{i, j=1}^{|V|} \sum_{e \in E} \frac{w(e)h(v_i, e)h(v_j, e)}{\delta(e)} \left\| \frac{\mathbf{f}_i}{\sqrt{d(v_i)}} - \frac{\mathbf{f}_j}{\sqrt{d(v_j)}} \right\|^2$$

$$+ \mu \sum_{i=1}^{|V|} \|\mathbf{f}_i - \mathbf{y}_i\|^2 \qquad (6.7)$$

It accumulates the changes in the score vectors between adjacent nodes over the hyperedges on the hypergraph. And μ is the regularization parameter which is greater than zero. The cost function should be minimized to keep the lowest loss, as shown in Eq. (6.8).

$$\min_{\mathbf{f}:V \to \mathbb{R}} Q(\mathbf{f}) \qquad (6.8)$$

One can derive the optimal solution of $\mathbf{f}$ by keeping the gradient of $Q(\mathbf{f})$ as zero, denoted as Eq. (6.9).

$$\frac{dQ}{d\mathbf{f}}\bigg|_{\mathbf{f}=\mathbf{f}^*} = (\mathbf{I} - \mathbf{A})\mathbf{f}^* + \mu(\mathbf{f}^* - \mathbf{y}) = 0 \qquad (6.9)$$

Then the optimal ranking score function $\mathbf{f}^*$ is derived, denoted as Eq. (6.10).

$$\mathbf{f}^* = (\mathbf{I} - \alpha\mathbf{A})^{-1}\mathbf{y} \qquad (6.10)$$

where $\alpha = 1/(\mu +1) \in (0, 1)$ is a parameter and $\mathbf{A} = \mathbf{D}_\mathbf{v}^{-1/2} \mathbf{H}\mathbf{W}\mathbf{D}_\mathbf{e}^{-1}\mathbf{H}^\mathbf{T}\mathbf{D}_\mathbf{v}^{-1/2}$ is an intermediate matrix.

The above process shows the deduction process of scoring function by direct derivation method, the optimal ranking score can also be derived via the random walks with restarts model on the hypergraph. Given a set of starting nodes in the vertex set V, expressed as a vector $\mathbf{q} \in \mathbb{R}^{|V|}$ where the elements of the starting nodes are 1 and other elements are 0, a random walk will transit to a neighbor node following an edge with probability α or restart from the starting node with probability $(1 - \alpha)$. We denote $\mathbf{p}^{(t)} \in$

$\mathbb{R}^{|V|}$ as the transition probability vector from a certain node to the other nodes at time t, and denote matrix $\mathbf{T} \in \mathbb{R}^{|V| \times |V|}$ as the transition matrix. The random walks on the graph is a recursive process that follows Eq. (6.11).

$$\mathbf{p}^{(t+1)} = \alpha \mathbf{T} \mathbf{p}^{(t)} + (1 - \alpha)\mathbf{q} \tag{6.11}$$

The transition probability vector will converge to a stationary distribution after an infinite number of jumps. At this time, $\mathbf{p}^{(c)} = \mathbf{p}^{(t+1)} = \mathbf{p}^{(t)}$ and Eq. (6.12) are satisfied

$$\mathbf{p}^{(c)} = (\mathbf{I} - \alpha \mathbf{T})^{-1}\mathbf{q} \tag{6.12}$$

where $\mathbf{T} = \mathbf{D}_v^{-1}\mathbf{H}\mathbf{W}\mathbf{D}_e^{-1}\mathbf{H}^{\mathsf{T}}$. $\mathbf{T}$ can be regarded as the normalized version of matrix $\mathbf{A}$. Note that Eq. (6.12) has same structure with Eq. (6.10), proving the equivalence between the random walk transitions' result and the direct derivation's result (Mao, Lu, Han, & Zhang, 2019; Tan et al., 2011).

By deducing the above equations, one can derive the scoring function via the random walks' convergence. Nevertheless, the original hypergraph ranking algorithm still cannot perfectly fit into the practical application of Smart PSS configuration. The hyperedge degree δ_e can vary dramatically, making each transition prone to the edge that has fewer nodes. This phenomenon is called "bias" on hypergraph. It often happens in Smart PSS configuration since some PSBs are high-end with lots of functional attributes (RAs) while others have fewer RAs. Meanwhile, it also happens when some users give detailed product comments while others just gave few words. The phenomenon can also be mathematically presented through the transition probability from vertex u to an adjacent vertex v in Eq. (6.13). The less the $\delta(e)$ is, the greater the $p(u, v)$ will be.

$$p(u, v) = \sum_{e \in E} w(e) \frac{h(u, e)}{d(u)} \frac{h(v, e)}{\delta(e)} \tag{6.13}$$

Nevertheless, the hyperedge degree should have less effect on the random walks since a hyperedge with more nodes does not equal to higher importance in the Smart PSS configuration.

To solve the bias on the hypergraph, an unbiased hypergraph ranking algorithm is adapted here by normalizing the transition probability, as defined in Eq. (6.14).

$$p(u, v)' = \sum_{e \in E} w(e) \frac{h(u, e)}{d(u)} \frac{\sqrt{\delta(e)}}{\sum_{e \in E} \delta(e)} \tag{6.14}$$

The modified transition matrix $\mathbf{T}'$ will be $\mathbf{D}_v^{-1}\mathbf{H}\mathbf{W}\mathbf{D}_e^{\frac{1}{2}}\left(\mathbf{M}^{\mathrm{T}}\mathbf{D}_e^{\frac{1}{2}}\mathbf{M}\right)^{-1}\mathbf{H}^{\mathrm{T}}$, where $\mathbf{M} \in \mathbb{R}^{|V|}$ is an associate column vector in which all the elements equal to 1.

6.4.3 Online application: user queries identification

Up to now, one has known how to learn the ranking scoring function based on the established hypergraph; the next step is to rank the possible PSBs when a user query is given online. This subsection will elaborate on the process of user query identification for the Smart PSS configuration.

As mentioned, both the functional RAs and the USs are applied in the user queries. A commonly applied user query identification method lets all the user-mentioned RAs and USs equal 1 in a query vector $\mathbf{q} \in \mathbb{R}^{|V|}$, and the other elements equal 0.

Although the earlier-mentioned method has already been proved to be effective for user query identification, the neighbor nodes of the user-selected nodes can also reflect the users' preferred RAs. Technically, after letting the initial query vector as $\mathbf{q}$ in which the user-mentioned entities as 1, we can obtain the final query vector as $\mathbf{y} = \mathbf{T}\mathbf{q}$. $\mathbf{T}_{\mathbf{u}, \mathbf{v}}$ represents the element of $\mathbf{T}$, meaning the transition probability from entity u to entity v.

As a result, the ranking scores of the PSBs can be calculated according to the final query vector $\mathbf{y}$ and the trained ranking function $\mathbf{f}$. The top K results can be recommended to the users.

6.5 Case study

To demonstrate how the established graphs can be used and to validate the feasibility of the algorithms, a case study of online 3D printing services was studied.

Today many 3D printer companies have launched online 3D printing services to the users, containing instant quotation, 3D model download and editing, remote 3D printing, postprocessing, and so on. These services can be flexibly integrated as plenty of PSBs, and they are distinguished by various 3D printing technologies, materials, printer models, and post-processing of the 3D printed parts. In this example, the users do not own a 3D printer but pay for the printing services, which makes it a typical result-oriented PSS (Tukker, 2004).

In the following subsections, the author will first illustrate the requirement elicitation process based on the established RG, and then show the solution configuration process based on the hypergraph.

6.5.1 Case study of the requirement management based on requirement graph

6.5.1.1 Requirement graph construction

The goal is to explore the implicit requirements that contain related components based on the given contexts in a 3D printing system. A dataset containing 1801 Q&As was collected from an online 3D printing discussion forum (https://forum.lulzbot.com/c/general/lulzbot-taz/13) where the users post questions and issues to find answers. After sorting and selecting typical issues/contexts, 622 Q&As were applied to construct the graph. The graph comprises 20 product components, six service modules and 145 contexts, such as "Nozzle," "LCD screen," "Parameter configuration," "adding extruder," and "adhesion issue," as shown in Fig. 6.8. In addition

Figure 6.8 3D printer-related components.

to the nodes, the constructed graph constitutes relationships between product components and contexts (P—C), service modules and contexts (S—C), product components and service modules (P—S), product components themselves (P—P) and service modules themselves (S—S).

The RG was constructed via a graph database, Neo4j, as shown in Fig. 6.9. In the graph, blue nodes with node names represent product component nodes. Similarly, orange nodes and red nodes represent service module nodes and context nodes, respectively. The graph shows that (1) many contexts exist in the 3D printing service system, meaning that the 3D printing service system is complex and the users face many different and uncertain situations when they use 3D printers; (2) the product nodes and service nodes are densely connected with other nodes, whereas the context nodes are sparsely connected because of the small number of product components and service modules; and (3) based on the accessible data, the contexts are not connected with each other, which is another reason for the sparsely connected context nodes. This phenomenon is also observed in the statistical information of the constructed graph, as shown in Table 6.3.

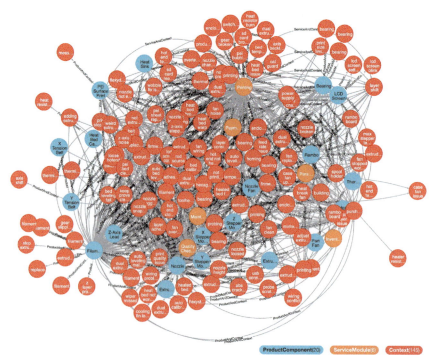

Figure 6.9 RG of the 3D printing service.

Table 6.3 The pseudocode of extracting similarities between nodes.

Element	Element name	Size	Average degree	Minimum degree	Maximum degree
P	Product components	20	24.15	1	148
S	Service components	6	41.3333333	2	183
C	Context	145	4.70547945	1	35
PC	Product-context	460			
SC	Service-context	227			
PS	Product-service	13			
PP	Product-product	5			
SS	Service-service	4			

Clearly, the average degree of product component nodes is 24.150, and the average degree of service module nodes is 41.333; nevertheless, the average degree of context nodes is only 4.705. The average degree of all the nodes is 8.2442. Due to the complex situations in the 3D printing service system, the issues can be caused by a series of components or parameter settings that the users cannot express precisely and truly, or cannot find the root causes. For example, the "heat bed not heating up" can be caused by either "wiper missed" or "thermistor issue," but the users can only describe the performance: not heating up. This issue conforms to the needs of extracting the aforementioned implicit requirements. Therefore, the sparse linkages of context nodes should be enriched to solve this problem.

6.5.1.2 Requirement elicitation

This section extracts the top K relevant nodes based on the given context node. During the process of obtaining node sequences via the random walk model, some parameters are set as follows. The constructed RG is treated as an undirected graph; all the edge weights are equal to 1, the window size of the random walk is 2, and the walk length is 40. By leveraging deepwalk, the 171 nodes with edges are embedded into a 16-dimensional space $\mathbb{R}^{16}$. The learned embeddings in $\mathbb{R}^{16}$ embrace the semantic relations among nodes, and they can also be easily used for computational operations. Finally, the cosine similarity between two nodes is computed to represent the similarity of nodes. The codes are implemented in Python and the scikit-learn package on Jupyter Notebook.

144   Smart Product-Service Systems

Table 6.4 Part of the extracted tuples.

Initial context nodes	Potential related nodes	Similarity
Nozzle clogged	Print quality issue	0.8662
	Extruder calibration	0.8447
	Keep probe clean	0.7935
	Measuring print bed temperature	0.7866
	Filament	0.7181
Layer shift	Hot end not heating up	0.8682
	Wiring conflict	0.8656
	Case fan	0.8501
	Case fan upgrade	0.8453
	Rambo board	0.8446
Extruder calibration	Print quality issue	0.9132
	Keep probe clean	0.8910
	Heat bed warp	0.8778
	pei sheet replacement	0.8730
	Filament	0.8447
Heat bed not working	Cooling fin loose	0.9722
	Wiper missed	0.9524
	Thermistor issue	0.9463
	Hot end change	0.9329
	Auto calibrate	0.9312

Given a certain context node, the approach can extract the top five related nodes as tuples. Tuples are ranked in terms of similarity from high to low. Serval results are shown in Table 6.4. For instance, the context node "nozzle clogged" is related to "print quality issue" with the highest similarity, i.e., 0.8662, followed by "extruder calibration", "keep probe clean," "measuring print bed temperature," and "filament." The result shows the implicit semantic relations among the given context, so it can provide the users more information to understand the current situation and identify potential causes.

Finally, the extracted tuples reveal the direct relationships and indirect relationships among nodes to help designers/engineers identify the problems hidden behind a large volume of heterogeneous contexts/products/services.

6.5.1.3 Results discussion

Precision and recall were used as the evaluation metrics to evaluate the performance of the link prediction algorithm. Based on the constructed graph, we randomly eliminated half of the edges so that the prediction task

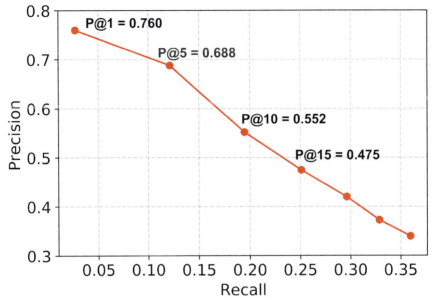

Figure 6.10 Precision-recall curve of the link prediction results.

is to find the missing edges via deepwalk. The results in Fig. 6.10 show that the precision is 0.76 when only one edge is predicted based on the given node. The precision decreases to 0.688 and 0.552 when the top 5 and top 10 related edges are predicted, respectively. Since the average node degree in the graph is only 8.2442, this result is feasible and acceptable.

The experimental results reveal that the proposed deepwalk-based requirement extraction approach can effectively eliminate latent user requirements while given a certain context. The proximity between nodes and their semantic relations are retained during the node embedding process, as demonstrated by the results with high similarity in Table 6.4.

6.5.2 Case study of solution configuration based on the hypergraph

6.5.2.1 Hypergraph construction

A dataset containing product specifications and product comments of the online 3D printing services was applied. From the dataset, the authors identified 28 PSBs, 219 RAs based on 24 kinds of parameters, and 2947 USs. Typical RAs include materials, nozzle number, category, 3D printer brand, application fields, printing speed, printing size, remote

146 Smart Product-Service Systems

Table 6.5 Statistical information of the hypergraph on 3D printing services.

Element	Element name	Size
PSB	Product-service bundle	28
RA	Requirement attribute	219
SC	Usage scenario	2947
$E^{(1)}$: PSB-RA	Product-service bundle-Requirement attribute relationship	28
$E^{(2)}$: PSB-SC	Product-service bundle-Usage scenario relationship	1714

printing model, additional functions and so on. Common USs include "high quality," "fast logistics," "cost-effective," "nondestructive packaging," "good customer service," "smooth surface," and so on. Here we have used a keyword extraction package on Python, viz. TextRank, on the product comments to identify the usage scenarios.

Based on the cleaned information, the hypergraph on 3D printing services can be established. Table 6.5 shows the statistics of the hypergraph. Note that the bias on the hypergraph can be clearly seen from the obvious difference between that the maximal edge degree (i.e., 20) and the minimal edge degree (i.e., 2).

6.5.2.2 Solution configuration

In this case study, we simulate the Smart PSS configuration as the process that a user identifies several RAs and SCs as the initial query, then the system will choose the top K PSBs as the results based on the ranking scores. The parameter K can be predefined by engineers in a case-by-case manner.

To evaluate the algorithm's performance, we treat the PSBs in the historical user reviews as the ground truth since they are exactly what the users selected. Totally 228 user reviews were randomly selected as the testing dataset, part of usage scenarios and RA were hidden to simulate the situation that the user will only care about partial but not all features of the PSB.

The hypergraph-based algorithm was conducted in the Python environment, together with a series of relevant packages. Particularly, NetworkX, and HALP (http://murali-group.github.io/halp/) were used to generate the hypergraph, and to conduct the PageRank algorithm and hypergraph ranking algorithm. Given the initial query vector $\mathbf{q}$, the final query vector was set as $\mathbf{y} = \mathbf{Tq}$ for the aim of enriching query vectors.

Table 6.6 Comparison with models.

	MAP	NDCG@3	NDCG@5	NDCG@8
PageRank	0.5260	0.4818	0.5195	0.5441
Hypergraph ranking	0.5567	0.5172	0.5516	0.5808
Modified hypergraph ranking	**0.5640**	**0.5493**	**0.6061**	**0.6201**

6.5.2.3 Results discussion

Two algorithms, namely PageRank and the original hypergraph ranking algorithm, are selected as the baselines to compare with the performance of the unbiased hypergraph ranking algorithm. We choose PageRank since it is a classical graph-based ranking algorithm who has good generalization capability.

Meanwhile, two evaluation metrics, i.e., mean average precision (MAP) and normalized discounted cumulative gain (NDCG), are applied. Table 6.6 shows the performance of these algorithms in terms of MAP and NDCG. The results show that the unbiased hypergraph ranking algorithm has the highest MAP as 0.5640, followed by the original hypergraph ranking. Meanwhile, it also has the best performance on NDCG whenever n equals to 3,5, or 8. The parameter K was set as small numbers since the total number of PSBs in this example is only 28, which is not large.

6.6 Chapter summary

Since Smart PSS is a complex system containing multipartite entities that their data formats are different, a compatible model is required to organize them. A graph-based model is an approximate description of the Smart PSS. Based on the graph-based models, an overall system architecture was proposed, including resource layer, heterogeneous data collection and storage layer, domain ontology construction layer, graph construction layer and application layer. In the graph construction layer, two graphs are defined:

(1) RG, containing product components, service components, and usage scenarios (i.e., usage contexts)

(2) Hypergraph for solution recommendation, containing RA, PSB, and usage scenarios

The requirement elicitation task in Smart PSS is formalized as a link prediction problem based on the given nodes. The relations between product components, service components and contexts are predicted, regarded as the implicit requirements. The results serve as references for engineers to

148 Smart Product-Service Systems

understand the potential requirements from users. The solution recommendation task in Smart PSS is defined as a ranking problem based on user queries. Specifically, both the technical requirements and the usage scenarios are allowed in the user queries. Graph-based algorithms including deepwalk and unbiased hypergraph ranking algorithm were applied. They are both compatible with the multipartite entities in Smart PSS.

References

Abowd, G. D., Dey, A. K., Brown, P. J., Davies, N., Smith, M., & Steggles, P. (1999). Towards a better understanding of context and context-awareness. In *International symposium on handheld and ubiquitous computing* (pp. 304—307). Springer.

Abramovici, M., Neubach, M., Schulze, M., & Spura, C. (2009). Metadata reference model for IPS2 lifecycle management. In *Proceedings of the 19th CIRP design conference—competitive design*. Cranfield University Press.

Amaxilatis, D., Mylonas, G., Diez, L., Theodoridis, E., Gutiérrez, V., & Muñoz, L. (2018). Managing pervasive sensing campaigns via an experimentation-as-a-service platform for smart cities. *Sensors, 18*(7), 2125. https://doi.org/10.3390/s18072125.

Barrett, K., & Power, R. (2003). State of the art: Context management. *M-Zones Research Programme,* 69—87.

Champiri, Z. D., Shahamiri, S. R., & Salim, S. S. B. (2015). A systematic review of scholar context-aware recommender systems. *Expert Systems with Applications, 42*(3), 1743—1758.

Chen, D., Chu, X., Su, Y., & Chu, D. (2014). A new conceptual design approach for context-aware product service system. In *2014 IEEE international conference on industrial engineering and engineering management* (pp. 1389—1393). IEEE.

Chen, G., & Kotz, D. (2000). *A survey of context-aware mobile computing research.*

Dey, A. K. (2001). *Understanding and using context. Personal and ubiquitous computing.* https://doi.org/10.1007/s007790170019.

Huang, K., Fan, Y., & Tan, W. (2014). Recommendation in an evolving service ecosystem based on network prediction. *IEEE Transactions on Automation Science and Engineering, 11*(3), 906—920.

Järvenpää, E., Siltala, N., Hylli, O., & Lanz, M. (2019). Implementation of capability matchmaking software facilitating faster production system design and reconfiguration planning. *Journal of Manufacturing Systems, 53*, 261—270.

Kim, S., Park, S., Lee, J., Jin, Y., Park, H. M., Chung, A., ... Choi, W. (2004). Sensible appliances: Applying context-awareness to appliance design. In *Personal and ubiquitous computing.* https://doi.org/10.1007/s00779-004-0276-9.

Kim, Y. S., Wang, E., Lee, S. W., & Cho, Y. C. (2009). A product-service system representation and its application in a concept design scenario. In *Proceedings of the 1st CIRP industrial product-service systems (IPS2) conference.* Cranfield University Press.

Li, Y., Chu, X., Chu, D., & Liu, Q. (2014). An integrated module partition approach for complex products and systems based on weighted complex networks. *International Journal of Production Research, 52*(15), 4608—4622.

Li, Y., Tao, F., Cheng, Y., Zhang, X., & Nee, A. Y. C. (2017). Complex networks in advanced manufacturing systems. *Journal of Manufacturing Systems, 43*, 409—421.

Long, H. J., Wang, L. Y., Shen, J., Wu, M. X., & Jiang, Z. B. (2013). Product service system configuration based on support vector machine considering customer perception. *International Journal of Production Research, 51*(18), 5450—5468.

Mao, M., Lu, J., Han, J., & Zhang, G. (2019). Multiobjective e-commerce recommendations based on hypergraph ranking. *Information Sciences, 471*, 269–287.

Schilit, B., Adams, N., & Want, R. (1994). Context-aware computing applications. In *1994 first workshop on mobile computing systems and applications* (pp. 85–90). IEEE.

Schmidt, A., Aidoo, K. A., Takaluoma, A., Tuomela, U., Van Laerhoven, K., & Van De Velde, W. (1999). Advanced interaction in context. In *Lecture notes in computer science (including subseries lecture notes in artificial intelligence and lecture notes in bioinformatics).* https://doi.org/10.1007/3-540-48157-5_10.

Stokic, D., & Correia, A. T. (2015). Context sensitive Web service engineering environment for product extensions in manufacturing industry. In *Service computation 2015* (pp. 9–13).

Tan, S., Bu, J., Chen, C., Xu, B., Wang, C., & He, X. (2011). Using rich social media information for music recommendation via hypergraph model. *ACM Transactions on Multimedia Computing, Communications, and Applications, 7*(1), 1–22.

Team, N. J. D. (n.d.). Tutorial: Import relational data into Neo4.

Teixeira, M. S., Maran, V., de Oliveira, J. P. M., Winter, M., & Machado, A. (2019). Situation-aware model for multi-objective decision making in ambient intelligence. *Applied Soft Computing*, 105532.

Tukker, A. (2004). Eight types of product–service system: Eight ways to sustainability? Experiences from SusProNet. *Business Strategy and the Environment, 13*(4), 246–260.

Uddin, M., Khaksar, W., & Torresen, J. (2018). Ambient sensors for elderly care and independent living: A survey. *Sensors, 18*(7), 2027.

Wang, Z., Chen, C.-H., Zheng, P., Li, X., & Khoo, L. P. (2019). A graph-based context-aware requirement elicitation approach in smart product-service systems. *International Journal of Production Research*, 1–17.

Wu, Z., Liao, J., Song, W., Mao, H., Huang, Z., Li, X., & Mao, H. (2018). Semantic hypergraph-based knowledge representation architecture for complex product development. *Computers in Industry, 100*, 43–56.

Zheng, P., Chen, C.-H., & Shang, S. (2019). Towards an automatic engineering change management in smart product-service systems—A DSM-based learning approach. *Advanced Engineering Informatics, 39*, 203–213.

Zhou, D., Huang, J., & Schölkopf, B. (2007). Learning with hypergraphs: Clustering, classification, and embedding. In *Advances in neural information processing systems* (pp. 1601–1608).

CHAPTER 7

Digital twin-enhanced product family design and optimization

Contents

7.1 Digital twin-enabled servitization	151
7.1.1 Engineering product family design and optimization	153
7.1.2 DT-driven product design and optimization	154
7.2 Trimodel-based generic framework	155
7.2.1 DT architecture featuring ambient information	156
7.2.2 In-context solution generation	158
7.3 DT-enhanced product family design	160
7.3.1 Benchmark mechanism for ambient-based product family design	161
7.3.2 Interacting mechanism for context-aware product family design	163
7.4 DT-driven product family optimization	165
7.4.1 DT-driven reconfiguration	167
7.4.2 DT-driven reverse design	167
7.5 Case study featuring a context-aware DT system	170
7.5.1 Establishing a context-aware DT system	170
7.5.2 Key advantages enabled via in-context solution design	174
7.6 Chapter summary	176
References	177

7.1 Digital twin-enabled servitization

With growing demand for higher product variety and better user–oriented services, industries are progressing toward the paradigm of mass personalization to gain competitive advantage (Tseng, Jiao, & Wang, 2010). Although manufacturers can gain market influence with increased product variety, rigid production processes prevent correlated profit increase due to the huge costs required to produce these customized goods. The situation, also known as the paradox of choice, is widely researched (Li, Lin, Chen, & Ma, 2015). Many existing product family approaches such as modular design (Dahmus, Gonzalez–Zugasti, & Otto, 2001), platform–based scalable design (Simpson, 2004), design structure matrix (Browning, 2016), and adaptable design with open architecture product (Zhang, Xue, & Gu, 2015) have exhibited effective results and are often utilized by industries. For

---

Smart Product-Service Systems
ISBN 978-0-323-85247-0
https://doi.org/10.1016/B978-0-323-85247-0.00003-7

© 2021 Elsevier Inc.
All rights reserved.

design processes, these approaches are typically conducted at the onset of the conceptual phase where mapping tools such as quality function deployment (QFD) translate functional requirements into design parameters. During the usage stage, reverse design occurs where essential feedback and information are collected to determine key parameters for improvement. These processes are tedious and time-consuming due to the lack of context-aware testbeds to support product family design and optimization with considerations to a holistic life cycle approach.

As enterprises embark on digitization roadmaps to realize Smart PSS that was brought forth by recent advancements in information and communication technologies (ICT), new paradigms have been proposed to highlight their effectiveness. An enabling technology, digital twin (DT), has emerged as a critical component in this digital servitization era. The concept of DT was originally coined by Grieves in a university presentation in 2003 (Grieves, 2011), and defined as a virtual representation of what has been produced for product life cycle management (PLM). According to Grieves (2014, pp.1−7), it mainly contains three parts: "(a) physical products in Real Space, (b) virtual products in Virtual Space, and (c) the connections of data and information that ties the virtual and real products together." Since then, several different definitions or expressions have been brought up. Nevertheless, the core characteristics maintain the same, and have been summarized by Tao et al. (2017) as (1) real-time reflection, (2) interaction and convergence in physical space, between historical data and real-time data, and physical and cyber space, and (3) self-evolution. They further point out nine aspects for DT-enabled service innovation in the manufacturing field: (1) real-time state monitoring, (2) energy consumption analysis and forecast, (3) user management and behavior analysis, (4) user operation guide, (5) intelligent optimization and update, (6) product failure analysis and prediction, (7) product maintenance strategy, (8) product virtual maintenance, and (9) product virtual operation. In literature, most existing studies of DT are implemented in the field of aeronautics and astronautics for maintenance purposes in the PLM, such as reducing maintenance cost (Tuegel, 2012), real-time inspection, repair and replacement (Hochhalter et al., 2014), prediction of failure (Cerrone, Hochhalter, Heber, & Ingraffea, 2014), etc. This is due to the fact that the DT enables a cost-effective simulation, monitoring, control, and prediction of the physical products, especially the ones with large complexity and high cost. Another fact is that the realization of high-fidelity, real-time simulation, and control of DT is enabled by CPS, which refers to the "tight conjoining of and coordination between computational and physical resources" (Baheti & Gill, 2011). As pointed out by Lee (2008, pp. 363–369), "embedded computers and networks monitor and control the physical processes, usually with feedback loops where physical processes affect computations and vice versa."

As a high-fidelity virtual replica along with real-time two-way connectivity, DT also leverages simulation and predictive technologies to provide decision-support features to aid stakeholders in fulfilling a wide variety of applications. With many existing DT studies conducted on PLM aspects, DT has the potential to achieve complex product family design and optimization with context awareness for Smart PSS development. As an explorative study, this chapter proposes a generic DT-driven approach with four mechanisms to support in-context solution generation for complex product family design and optimization.

7.1.1 Engineering product family design and optimization

Product family design has evolved over the years to provide personalized products to fulfill consumer demands. While manufacturers can gain market influence with increased product varieties, rigid production processes prevent correlated profit increase due to the huge costs required to produce these customized goods. As such, the product family approach offers a cost-effective solution and considers customer requirements in both functional and physical domains. Modular design defines and selects optimal design solutions while scalable design improves asset configurations to enhance existing solutions and aid design co-creation. Simpson, Jiao, Siddique, and Höltta-Otto (2014) defines product family design as a product cluster with a network of interchangeable modules for a range of functionalities to satisfy various market demands. As a key concept within the mass customization paradigm, product family solutions are often developed with business-driven considerations in mind. Studies to increase product family design capabilities have also been conducted with Zheng, Xu, Yu, and Liu (2017) presenting a platform-based adaptable design while a cost-effective design approach for additive manufacturing was featured by Yao, Moon, and Bi (2016). Likewise, other relevant studies from Simpson et al. (2014) and Jiao, Simpson, and Siddique (2007) can be looked into as additional references. With increasing research interest in technology-driven solutions, Ma and Kim (2016) showcased a data-driven clustering/mobile partition approach utilizing AI techniques to perform preference and function prediction made possible with online data availability. Additionally, a smart grid service was designed based on lifestyle details from smart appliance logs (Takenaka, Yamamoto, Fukuda, Kimura, & Ueda, 2016) and the use of a cyber-physical codesign approach by Zheng, Xu, & Chen (2018) to support smart product family design. Although these studies demonstrate the advancement in family solution design to fit today's context of smart environment applications, the mapping processes utilized lack visualization and do not actively consider ambient information during the design process.

Conversely, product family optimization typically occurs in the usage stage and aims to enhance existing asset families (D'Souza & Simpson, 2003). Quite often involving engineering management processes such as redesign management, product family optimization approaches such as design structure matrix (DSM) allows the identification of links between entities (Qiao, Efatmaneshnik, Ryan, & Shoval, 2017). Alternatively, heuristic and bioinfluenced processes such as ant colony optimization by Wei, Tian, Peng, Liu, and Zhang (2019) can achieve redesign efficiency by minimizing design propagations. The incorporation of smart enabling technologies by Zheng, Chen, and Shang (2019) introduces a DSM-based learning while Savarino, Abramovici, Göbel, and Gebus (2018) showcases a smart reconfiguration approach to loop design iteration cycles. Product family optimization can also provide collective improvements to asset configurations for reuse and remanufacturing processes. For instance, a Stackelberg game theory-based approach was proposed by D. Wang et al. (2016) to optimize modular and scalable designs. A joint decision-making approach for product family optimization was utilized by Wu, Kwong, Lee, and Tang (2016) with considerations to both design and remanufacturing aspects. As such, the integration of technologies such as DT technologies can provide process visualization capabilities while the use of ambient information can aid designers in enhancing solution redesign and reconfiguration.

7.1.2 DT-driven product design and optimization

In recent years, DT has emerged as a trending technology due to rapidly evolving simulation and modeling capabilities, better interoperability, and low-cost IoT sensors. First introduced in 2003 (Grieves, 2014, pp. 1—7), DT has evolved over the years, integrating high-fidelity CAD models, knowledge representation, and computational tools across a growing range of industries such as construction, healthcare, and even horticulture. While DT offers an inexpensive approach of investigating and evaluating complex systems, there is a lack of a unified definition and is often regarded as a high-fidelity virtual replica with real-time two-way communication for simulation purposes and decision-support capabilities for product-service enhancement. As many existing DT studies consider PLM aspects (Lim, Zheng, & Chen, 2019; Tao et al., 2017), this study focuses on design and inverse design-related work. In the geometry assurance domain, Schleich, Anwer, Mathieu, and Wartzack (2017), applied the Skin Model Shapes concept for design geometry inspection via a holistic DT reference model while Biancolini and Cella (2018) established a DT-driven mesh morphing workflow for geometric model validation. In the value co-creation domain, Zheng et al. (2018) presented a data-driven DT platform to support value co-creation within smart connected environments. Zheng, Lin, Chen, & Xu (2018) also featured a codesign methodology for the development of smart wearables using DT technologies. In the model-based design and simulation domain, DT is incorporated into model-based systems to conduct context-aware simulations (Schluse, Priggemeyer, Atorf, & Rossmann, 2018). Similarly, Damjanovic-Behrendt and Behrendt (2019) created a DT demonstrator for high-fidelity product simulation via open-sourced resources.

While the aforementioned literature has demonstrated DT's effectiveness enhancing design and optimization processes, these DT applications only serve a single PLM aspect and the lack of context-aware capabilities resulted in rigid DT solutions. Moreover, the strong emphasis on computational approaches rather than on modeling processes in these studies meant that there is minimal work on realistic testing approaches for product evaluation. As such, this study aims to develop context-aware DT systems to facilitate generative design, integrate ambient information to aid both in-context solution design and redesign in a systematic and holistic manner via a tri model framework.

7.2 Trimodel-based generic framework

To fulfill the challenges highlighted previously in Chapter 7.1, a generic DT system with context-aware capabilities is presented featuring the integration of ambient information. This DT system is based on a four-layered cyber-physical architecture, where key features such as high-fidelity simulation, real-time two-way communication, and data-driven decision-support functions form essential components for expediting design processes. To achieve a functional DT for complex product family design, the work follows the hierarchical data-information-knowledge-wisdom

(DIKW) model (Rowley, 2007), which leverages on product-related data modeling and mapping to achieve design efficiency. This generic architecture aims to provide a low-cost and effective solution to expedite the design and planning phase as well as on-site usage phase. Chapter 7.2 features key techniques toward integrating two-way communication, simulation, and decision-support capabilities to establish a functional DT system aimed at tackling design and usage inefficiencies brought forth by complex environments. To achieve in-context solution design and workflow optimization, this study takes reference from a trimodel DT framework (Zheng & Shankar, 2020) described in the subsequent chapters. Leveraging real-time context monitoring of ambient events and featuring the use of product family assets in dynamic environments, this proposed DT model is also applied to a case study in Chapter 7.5 to highlight its effectiveness and capabilities.

7.2.1 DT architecture featuring ambient information

Chapter 7.2.1 proposes a DT architecture featuring a trimodel generic DT architecture to generate in-context solutions. Starting with Fig. 7.1, five essential DT components are integrated with arrows highlighting the data

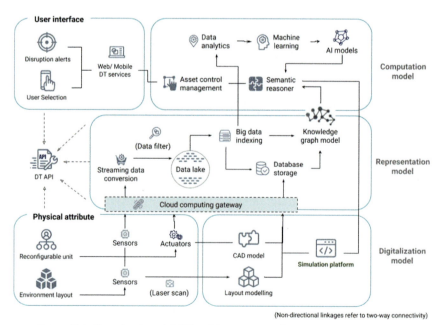

Figure 7.1 Generic trimodel DT architecture to achieve context awareness.

flow within the model. Starting from the physical attribute component, the goal is to establish a two-way communication flow between the physical attributes, which consists of both the product family asset and its current environment layout, and the digitalization model. Sensors are embedded at strategic locations on both asset modules and environment layout to obtain relevant data, which is transferred via industrial communication protocols (e.g., OPC-UA and MQTT) to achieve consistent and reliable data transfer. As for the digitization model, the modeling process begins with the creation of high-fidelity CAD models for all product family modules while the environment layout is created either by using laser scan technology or 3D modeling software. Dimension accuracy is a major parameter and must not be compromised in the creation of the digitalization model.

With both the cyber asset and layout components mapped to reflect the real-time status of their physical counterparts, the arrows indicate data flow to the representation model via a cloud computing gateway. This gateway manages heterogeneous data pass on by IoT networks for data processing in the streaming data conversion component. In this module, the irrelevant sensor data is filtered out before temporary storage in the data lake. These data are then indexed in a data repository according to predefined categories where data can be extracted to perform data analysis on subjects such as condition monitoring. Both the layout and CAD models from the digitalization model as well as useful sensor data that is retained by users are ultimately stored in NoSQL databases due to their ability to handle large amounts of heterogeneous data. Both the database and big data indexing components perform storage and retrieval roles to ensure information transfer in a reliable and consistent manner. The information is then passed on to a preestablished ontology model where key parameters are translated as entities and relationships.

In the computation model, information is translated into valuable insights via knowledge graphs and optimization algorithms to enhance existing workflows. The ontology model, which includes stakeholder requirements, safety rules, and operation details, is then populated with real-time or forecasted data to create an effective knowledge graph. From the semantic reasoner, solutions inferred from this graph model are evaluated via a multiphysics model simulation to verify their feasibility. Data analytics and machine learning tools enable key statistics to be derived, providing operators with transparent solutions to make informed decisions. These potential solutions are ranked and presented to end users in a comprehensive layout and features an interface for asset control. The user interface component forms the final aspect of the DT architecture featuring web/mobile-based services for users to control and monitor product family assets. Capable of detecting disruptions stemming from the physical attributes, this DT system is also capable of in-context solution generation for disruption management and process optimization while facilitating asset re-/configuration processes.

Due to intense data applications required by each model, it is crucial to maintain seamless data transactions between each model. To ensure reliable and consistent data transactions, industrial machine-to-machine communication protocols such as MQTT and OPC-UA are utilized in accordance with Industry 4.0 frameworks. The reference architecture model industrie 4.0 (RAMI 4.0) (Moghaddam, Cadavid, Kenley, & Deshmukh, 2018) is an interdisciplinary framework suitable for next generation product development with the incorporation of business models. Data packets retrieved from the physical attribute component are processed and converted into machine-readable format for temporary storage in the data lake. OPC-UA, being a cross-platform industry standard, supports both SOAP (web service protocol) and binary protocol which enables smooth data transition through firewalls. Based on the seven-layered open systems interconnection (OSI) model, encrypted information is passed to subsequent components within the DT architecture efficiently. Alternatively, disruption and control signals from the user interface will be converted into small data packets and forwarded back to the physical attribute component through the OSI layers with the same machine-readable format. Reference libraries and rules within the APIs also enable different software to be integrated, allowing external programs access to information while acting as a control center for each component.

The generic architecture proposed in this study is highly versatile as compared to those from DT-related studies due to the use of a trimodel-based approach to achieve context awareness. By leveraging cognitive intelligence, this approach enables better synergy with physical attributes than conventional geometrical design–based DT to enable both monitoring and control capabilities.

7.2.2 In-context solution generation

In-context solution generation forms the crux of the context-aware DT system and allows stakeholders to make informed decisions in a shorter amount of time. Condensed into a process flow diagram for clarity, Fig. 7.2 depicts the computational logic which begins from the physical attribute to

Digital twin-enhanced product family design and optimization 159

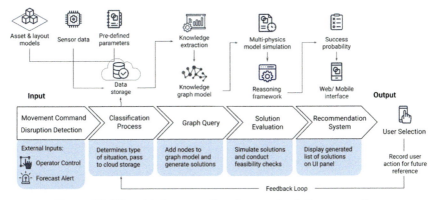

Figure 7.2 DT-enabled process flow for smart solution generation.

the user interface component. The DT system is triggered when user input or disruption alerts are detected, initiating a classification process to determine the activity type depending on the exceeded parameter thresholds. The information is then transferred via cloud platforms to the knowledge extraction module where it is translated into event entities for input into the knowledge graph model. Solutions related to product family optimization and disruption management are then inferred via graph queries and verified using a multiphysics model simulation to reject unviable options. These solutions then go through a process of elimination based on predefined factors such as time-cost-space constraints, functionalities, and user preferences. The remaining solutions are ranked according to predefined priorities and displayed via the web/mobile interface. The system also keeps track of user selections to perform feedback loops for knowledge reuse in similar situations.

This DT-enabled process flow aims to provide transparency with regards to solution generation so that stakeholders can make informed decisions and facilitate tweaks to the logic flow for alternate use cases. The system serves as a fundamental for smart automated workflows designed to manage multiple assets and deal with a variety of disruption cases. The level of automation can also be tweaked to include manual control elements for high-risk operations while regular processes can be fully autonomous with human supervision. Over time, user confidence in the DT system would increase due to its transparency and flexibility in aiding stakeholders to make informed decisions, thus freeing up valuable resources during the usage stage.

7.3 DT-enhanced product family design

While the aforementioned generic architecture provides a process model to recognize the data flow going between the DT components, Chapter 7.3 provides an insight into how a context-aware DT system is able to aid product family design and optimization (Lim, Zheng, Chen, & Huang, 2020) with considerations to Smart PSS attributes. This section further takes reference from the cyber-physical system highlighted in the previous architecture, and proposes two mechanisms to support product configuration and evaluation in the planning phase. Every initial project commencement starts with the design phase in which the customer places requests for stakeholders such as architects and designers to plan and implement an optimal product solution designed to tackle pain points. While typical processes within the industry relies on third-parties to survey the site and provide recommendations based on experience, these steps are not only costly and time-consuming, but also prone to oversight and lapses in judgment, resulting in reverberating consequences affecting project efficiency and product utilization rates. This approach allows a quicker and economic manner of performing such tasks with the system that can generate solutions for multiple scenarios, making it reusable for similar operations.

Adopting the product family approach, Fig. 7.3 showcases an overview of the design and planning process to identify optimal product configurations incorporating ambient information to allow better visualization as to how potential customized products will perform in the intended environment. Based on the initial customer expectations and project requirements, these information are imported into the preestablished knowledge graph known as constraint entities. These constraints include time and budget stipulations and are imposed as rules during the solution generation process followed by obtaining context information from the target scenario from both static and inconstant parameters.

To map modular product components from a preferred range of families, high-fidelity CAD models of product components with accurate dimensions as well as their relevant functionalities are stored in the same knowledge graph to infer optimal configurations. After deriving the potential configurations, these compatible modules are combined and loaded onto the common simulation platform to visualize and further test the functionalities and effectiveness of the product. After this evaluation phase, a final product configuration optimal for the current situation would be

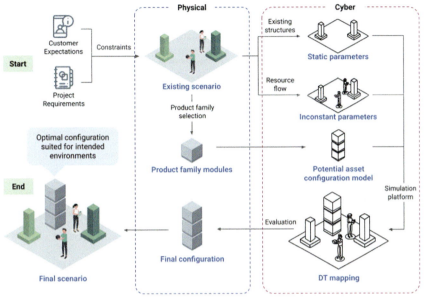

Figure 7.3 Cyber-physical system with ambient information input for product family design.

identified, with the entire design and planning process taking less than a day as compared to weeks when outsourced to third-party experts. Moreover, additional project requirements could be factored in at any point to support the co-creation process, providing more value for stakeholders without compromising on time and quality. This would be highly beneficial for industrial usage, especially in manufacturing and construction sectors where end users can assess product performance instead of solely relying on expert's opinions. Chapters 7.3.1 and 7.3.2 dive further into the inner workings of this design process with the use of benchmark and interaction mechanisms with Smart PSS elements taken into consideration.

7.3.1 Benchmark mechanism for ambient-based product family design

The benchmark mechanism is designed to assist in the initial product configuration process as well as identification of suitable locations for installation. Depicted in Fig. 7.4, the modeling process starts with mapping of the intended physical environment layout onto a virtual environment. This initial process focuses on identifying all existing static assets and their physical movement ranges. To map out the static parameters, which refers

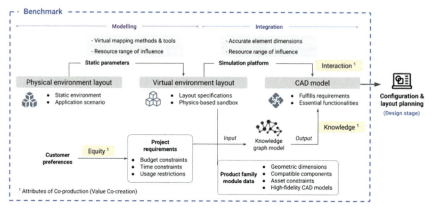

Figure 7.4 Benchmark mechanism for product family configuration and layout planning.

to existing structures and large equipment, point-cloud technology can be used to map out the entire work site or shop floor into a 3D environment. Alternatively, CAD models can also be uploaded with emphasis on dimensions depicting accurate range of activities. The next process involves integrating both the virtual environment layout and the virtual product model onto a common simulation platform. To identify potential product configurations, customer preferences are elicited and factored as a part of the project requirements and constraints. These attributes typically relate to the project duration and implementation costs while other regulatory guidelines and layout restrictions (especially in shop-floor scenarios) should be factored in to derive suitable recommendations. The generation of optimal product configurations requires a wide range of family module data ranging from functionalities to physical constraints that are fetched from a dataset and inserted into the knowledge graph model.

The knowledge graph model functions as described previously in Chapter 7.2.2 and infers several configuration recommendations based on the predefined constraints. These product configurations are composed of compatible modules that are then further evaluated in the virtual environment. This proposed benchmark mechanism ultimately incorporates coproduction features, a key concept in the Smart PSS paradigm. Falling under the notion of value co-creation (Ranjan & Read, 2016), coproduction consists of equity, knowledge, and interaction attributes. Equity enables end users and designers to contribute toward the design process by facilitating the collaboration and consolidation of ideas to further advance mutual interest and goals. Reflected in the project requirements, the aim of equity would result in the development of meaningful products and actualize key concerns by both parties. The knowledge attribute refers to knowledge sharing which integrates datasets from product family modules, ambient sources, and predefined constraints and preferences to provide better outcomes to address dynamic concerns. This is shown using knowledge graphs to link the relationships between various entities, allowing designers to consider key factors while maintaining a clear overview during decision-making processes. Lastly, the interaction attribute is realized by the simulation platform, providing an avenue to support stakeholders in understanding the challenges determined by the current static situation. The 3D models of both environment and potential product configuration offer both clarity and flexibility during the selection of suitable installation locations as designers can make repeated adjustments and alter product parameters to obtain the desired outcome. Besides coproduction, value co-creation also consists of the value-in-use concept which is realized in the following chapter featuring the interacting mechanism.

7.3.2 Interacting mechanism for context-aware product family design

While the benchmark mechanism forms an initial approach to outline and shape potential products, it is through the interaction mechanism that puts forward a feasible product family design by considering the ambient context. As shown in Fig. 7.5, this secondary process starts with the physical modular unit where sensors are embedded onto each modular component to monitor its status as well as surrounding context. With data supplied from each product module, designers will be able to determine the rigidity of the configuration under the intended working conditions. This not only sets the stage for product condition monitoring functionalities during the usage stage, but more importantly, provides a clear comparison between the potential configurations during the design stage. Inconstant parameters serve to provide real-time data related to present layout status such as resource movement and environment conditions. These parameters can be obtained by embedding sensors at strategic locations and placing RFID tags on moving assets such as manpower and AGVs. These heterogeneous data serve as a continuous source of information and are transferred into the knowledge graph and simulation platform to provide real-time visualization of both asset and environment layout conditions.

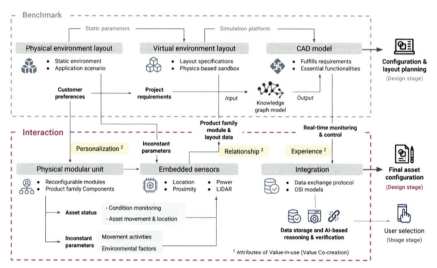

Figure 7.5 Interaction mechanism to evaluate potential product family configurations.

The simulation platform features two-way real-time communication capabilities and allows designers to monitor and control potential product models to evaluate performance indexes such as utilization rates and efficiency while identifying possible bottleneck processes. The evaluation of multiple configurations and iterations can be carried out without affecting current site operations and offer an unambiguous approach toward the final product. Data integration between the different mechanisms strictly follows industrial dataflow standards such as the 7-layered OSI framework and communication protocols such as OPC-UA and MQTT. The integration component consists of data architecture selection as well as data processing and management. Due to the exponential amount of sensor data generated over time, considerations have to be made to store only specific sets of information with intrinsic value using filtering rules and other methods. With the physical modules, sensors, and integration components in place, designers will be able to conduct simulation trial runs of prototype usage under varying conditions. This facilitates extensive testing with realistic use case scenarios that not only determines its durability but also its compatibility in conjunction with other equipment to avoid functionality mismatches or bottleneck situations.

The simulation platform provides a comprehensive tool to determine final product configuration and placement within customized scenarios and can be expanded to include other product-type families as well. Likewise, the platform is designed for reusability with the twin model providing support throughout the product life cycle stages, especially during the usage stage with condition monitoring and maintenance support. The interaction mechanism enables the establishment of a context-aware DT system with considerations to Smart PSS elements such as value-in-use, a key feature of value co-creation. Value-in-use focusing on acquiring quality and desirability through established processes and through the implementation of a DT system, can achieve personalization, relationship, and experience attributes in the design stage. Personalization allows customers to gain immersion into the design process based on preferences. Using product family modules allows flexibility in designing assets with unique values and takes into account parameters which are difficult to be classified as requirements. The relationship attribute allows users to have dynamic interactions with the virtual models and create value by considering ambient parameters during the design process. Furthermore, monitoring and control capabilities can build cooperative relations between designers and operators to better handle and evaluate the effectiveness of the product. Finally, the experience attribute provides intrinsic value by allowing designers to experience the various product functionalities to identify optimal configurations based on the intended use cases. Insights can also be obtained based on operator behaviors during the usage stage to improve usability, thus creating value for end users.

7.4 DT-driven product family optimization

While many DT-related studies have sought to tackle challenges faced in every PLM aspect ranging from the design to end-of-life phase, most of the existing DT solutions are limited to only a single PLM phase, thus resulting in rigid systems which are unable to continuously shadow the asset throughout its product life cycle. Using context-aware DT systems, both design and usage stages can be considered for a specific product, covering key factors to ensure that customers have better user experiences. This chapter focuses on the usage stage and highlights the use of ambient information to improve existing product family asset capabilities in the event of disruptions via a reconfiguration and redesign approach. This optimization approach allows stakeholders to maximize asset utilization rates with increased workflow efficiency as well as resolve bottleneck issues, while reducing failure rate throughout operation.

Fig. 7.6 showcases a process overview utilizing context-aware DT systems to resolve challenges commonly faced during product applications.

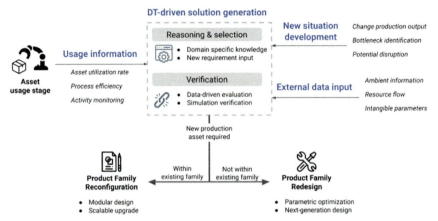

Figure 7.6 Reconfiguration and redesign process overview for product family optimization.

These challenges range from process automation such as path planning to production output switches. While studies on "conventional" DT approaches can boost relevant performance indexes such as utilization rate and workflow efficiency, context-aware DT systems can provide relevant support and value to stakeholders in specific scenarios. In this chapter, the use of "conventional" is noted as DT systems that do not require ambient information input and are typically focused on a singular asset and life cycle aspect.

The proposed DT process flow is capable of generating in-context solution design through utilizing ambient information to provide optimal decision-support for operators to ensure workflow continuity and reduce downtime risks. As such, DT systems carrying out condition monitoring and predictive maintenance duties can feature additional information from ambient sources to generate smart recommendations. For instance, a production asset facing a shortage in a particular material would feature a time-scheduled reminder instead of critical warning to the operator if subsequent batches of output do not require that material. This would free up resources which could be directed to urgent cases as the DT system would consider the asset as part of the wider working system. With usage information derived from the product and external data input from the developing circumstance, when a new situation arises, the DT-driven system would first infer a range of potential solutions and rank them accordingly. Next, simulations-led verifications are performed to evaluate feasible solutions depending on stakeholder interests and predefined constraints. If a potential solution requires an asset change, the DT system would consequently facilitate either the product family reconfiguration or redesign process. The selection of either process is determined by the existing product configuration and if the recommended solution falls within the same product family and are examined in the subsequent chapters.

7.4.1 DT-driven reconfiguration

The product family reconfiguration process aims to exchange compatible family modules to attain new functionalities to meet new product requirements. Based on information obtained from asset usage and ambient information gathered from sensors and IoT systems, new requirements elicited from disruptions are first categorized into a set of constraints and preferences. Next, the information sets are converted into entities which are entered into the knowledge graph model. Also known as requirements elicitation, the updated knowledge graph would then provide in-context solutions which are then compared with input from the crowdsourcing aspect. Crowdsourcing, being part of the Smart PSS paradigm, enables the use of collective wisdom based on user understanding and experience to influence the ranking of the recommendations presented to stakeholders. The advantages brought forth by crowdsourcing in the domain of product family optimization includes faster and superior decision-support systems. While crowdsourced information may include recurring configuration selections, new applications, and common scenarios, they are categorized by performance, dataset types, algorithms, and utilization. Sources vary from public and organizational input, providing reliability and consistency to the context-aware DT system. For instance, a supply chain-related disruption would redirect factories to begin mass production of a particular product. This decision would require the reconfiguration of existing production equipment whereby the DT system would provide in-context recommendations for asset relocation or module replacement to gain functionalities that increase production output. Fig. 7.7 showcases a streamlined representation of the reconfiguration process flow.

7.4.2 DT-driven reverse design

This chapter further elaborates on the product family optimization process with regards to the reverse design aspect in the event whereby DT systems are unable to determine a suitable product module for exchange. Following the same initial process as mentioned above, the new project requirements are translated into potential product attributes which are taken into

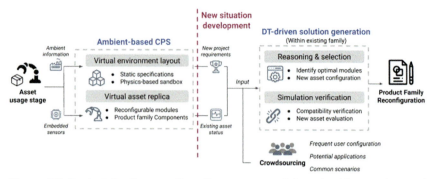

Figure 7.7 Product family reconfiguration process reinforced with crowdsourced insights.

consideration when designing next generation product family modules. These attributes, together with conceptual and prototype models from commercial and open-sourced designs, are uploaded into the knowledge graph model to further analyze key parameters. The output would feature new design parameters such as dimensions, manufacturing method, and cost constraints, which are essential toward product creation and innovation. Aided by generative design software such as CATIA by Dassault Systèmes and Creo by PTC, the derived parameters can be processed using a set of logic operations to produce a model for simulation-led evaluations. As such, this product family redesign approach facilitates new module generation to suit a wide range of applications. Incentive mechanism, a key aspect of the Smart PSS paradigm, aims to motivate external parties to contribute toward new product creation. To incentivize external parties to contribute toward the redesign process, a series of solutions ranging from simulation-verified algorithms, motivation styles, and an online-offline setting can be considered. The simulation-verified algorithms leverage stochastic models to determine the effect of randomness for participants to deal with related situations while categorical motivation refers to the type of incentives most appealing to participants. Such incentives can be in the form of monetary, entertainment, and service, thus fulfilling the needs of participants. Lastly, the online-offline settings would provide a setup to obtain necessary information via a simplified or comprehensive manner. Fig. 7.8 showcases the process flow featuring the product family redesign alternative for usage optimization.

To maximize asset availability and utilization rates, the development of context-aware DT enables the creation of a compelling decision-support

Digital twin-enhanced product family design and optimization   169

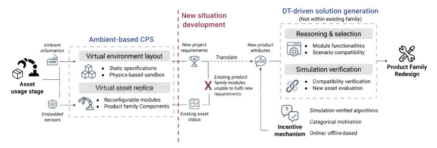

Figure 7.8 Product family redesign process coupled with incentive mechanisms.

system to assist shop-floor managers in achieving workflow efficiency. In addition, DT is capable of further evaluating resource flow with increased accuracy to generate viable solutions. These data-driven solutions facilitate planning operations in advance and prevent safety oversights, providing an efficient approach toward remote asset management. As such, this DT-enabled platform forms a foundation toward autonomous production with minimal human supervision in dynamic environments that are manpower-reliant.

As frequent disruptions occurring within these dynamic environments often result in work order delays and other adverse repercussions, there is a pressing need to utilize context-aware DT systems to avoid sudden setbacks by forecasting demand based on historical trends and simulating potential effects. These disruptions range from equipment failures to market factors affecting demand and through the use of real-time sensors and semantic monitoring tools, the DT computational component can provide simulation-evaluated solutions in addition to recovery measures such as production rescheduling while incorporating business model adaptations to provide stakeholders with a robust decision-support system to effectively overcome disruptions. Predefined time and cost priorities can be established based on stakeholder expectations and resources will be diverted to maximize existing outcomes. These in-context solutions enable labor-intensive operations with increased flexibility to overcome major supply disruptions and in some circumstances, assist planners with shop-floor layout conversions.

The benchmark and interaction mechanisms proposed are implemented in an industrial use case in the subsequent chapter to showcase both design and optimization capabilities. By leveraging on ambient information to actualize context awareness, a DT-enhanced service environment can be established as a low-cost solution to facilitate decision-support within

170 Smart Product-Service Systems

dynamic environments. Furthermore, the use of DT technologies as a service allows these smart systems to resolve major obstacles encountered throughout multiple PLM stages.

7.5 Case study featuring a context-aware DT system

Chapter 7.5 features a systematic approach toward the creation of a context-aware DT system which incorporates all the above-mentioned design and optimization methodologies. A detailed analysis on the fundamental DT technology drivers are discussed along with potential enabling tools and industry standards to provide a comprehensive assessment of the system's capabilities. In addition, this generic framework is supported with a case study featuring the use of tower crane DT service (Zheng & Lim, 2020). Emphasizing on the importance of utilizing ambient information within DT systems to facilitate product family approaches, this study also paves the way toward the development of a reusable support system which encompasses the entire asset life cycle and reshapes the way inefficiencies in projects are resolved. Lastly, in relation to the featured case study, a trimodel-based approach consisting of digital, computation, and representation models is introduced with considerations to expediting workflows in a cost-effective manner. The goal of this trimodel DT system ultimately serves to derive smart recommendations to mitigate or resolve an extensive range of problems within dynamic environments.

7.5.1 Establishing a context-aware DT system

The objectives of establishing a tower crane DT system based on the proposed framework mentioned in Chapter 7.2 can be narrowed down to three key functionalities. Firstly, efficient 3D modeling of customized tower crane parts. Secondly, visualizing and comparing tower crane configurations under various application scenarios. Thirdly, utilizing domain knowledge to perform design optimization. Using the trimodel approach, a DT-enhanced smart solution is proposed to expedite the planning and usage phase based on the tower crane DT as an example. The trimodel approach leverages the digital model for integration of high-fidelity product family asset and environment layout models within a simulation platform, representation model for determining relationships between parameters for solution inference, and evaluation model to facilitate decision-support systems by validating solutions designed for asset re-/configuration with the aid of both benchmark and interaction mechanisms.

In this case study, the digital model is categorized into asset and environment layout. Asset modeling serves as a vital step toward the creation of all DT-related systems and the product family approach is a viable technique to create digital replicas rapidly. As the conventional modeling approach involves modeling-compatible physical assets modules from scratch, this tedious and time-consuming process poses a major obstacle for rapid industry adoption. Alternatively, the product family approach allows both the asset-environment modules and relevant ontology schemas to be reused, thus expediting the CAD modeling processes and forming a fundamental toward DT establishment. The tower crane modules are modeled based on the Liebherr manufacturing blueprints, which includes the load range, maximum height, and precise dimensions. The modeling software, 3DsMax, is used to create this set of product family modules together with its functionalities such as movement, rotation, and range. These asset models are constructed using existing product family methodologies such as the top-down and bottom-up approaches. Comparison between different configurations is such that tower cranes within the same family would share the same components. The static layout model of an oil and gas site is established using the prevailing industrial software, plant design management system (PDMS). Alternatively, the environment model can also be created via point-cloud mapping of existing sites. Both the asset configuration and layout models are then uploaded onto a custom-made simulation system. With the existing structural requirements known, identification of suitable areas for tower crane placement can take place based on the construction requirements.

As existing digital modeling and simulation applications do not support modifications such as user input commands or ambient information input for creation of a context-aware DT system, a custom-made simulation platform is established to fulfill the case study's objectives. This platform is created using C++ and relies on libraries such as MFC and OpenGL to achieve specific DT requirements. Fig. 7.9 showcases the simulation platform with ambient information input on the left. Utilizing this platform, designers would be able to conduct benchmark and interaction tasks to plan and evaluate potential asset configurations with the environment.

The subsequent goal would be to establish relationships between entities which are categorized according to predefined parameters and project requirements via an ontology model. This model would essentially hold information relating to asset module specifications, layout boundaries and restrictions, stakeholder requirements, sensor input details, and many more.

172 Smart Product-Service Systems

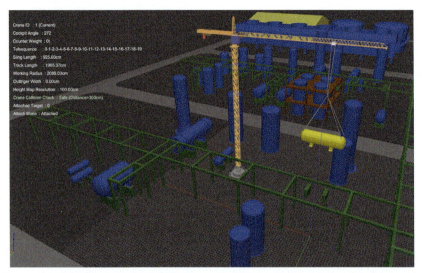

Figure 7.9 Simulation system featuring a functional tower crane asset and static environment layout.

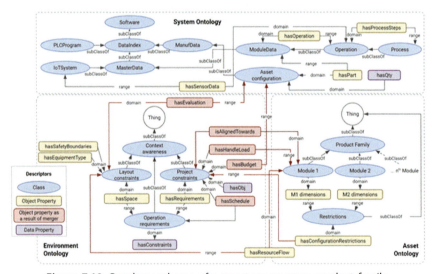

Figure 7.10 Ontology schema of a smart tower crane product family.

Fig. 7.10 depicts a simplified ontology of a smart tower crane product family. Additional ontologies such as sensor type and crane specifications can be created separately and added to the existing ontology, with relations specified to provide a dynamic reusable system which can adapt to different types of building scenarios.

The entities, attributes, and their interrelationships provide the domain-specific knowledge of the tower crane product family. Project specific ontologies consisting of customer expectations can also be mapped onto an ontology framework via a requirement elicitation approach (Wang, Chen, Zheng, Li, & Khoo, 2019). The acquisition of real-time quality sensor data can be integrated into the ontology framework via the Neo4j graph database to establish a knowledge graph. As such, a range of in-context solutions can be derived to expedite design and optimization workflows. This tower crane ontology consists of interactions between asset, environment, and other schemas such as project objectives and constraints to realize knowledge inference. Modeled after a learning pedagogy proposed by David et al. which is aimed at facilitating a learning environment for asset designing and optimization (David, Lobov, & Lanz, 2019), this enables optimal asset configurations suitable for the predefined layout to be generated and the constant stream of data input further enhances future asset PLM reconfiguration and process optimization. The proposed configurations can be evaluated via simulations as mentioned earlier. Capable of information storage and linking relationships, the graph-based model can achieve context-aware cognitions to fulfill product family design and optimization. When integrated with different AI models embedded in the DT computational layer, predictive capabilities can be enabled for maintenance, usage optimization, asset re-/configuration, and layout planning.

The components within this context-aware DT are connected via APIs which can be categorized into modeling, message broker, and engagement attributes (Fig. 7.11). The modeling API sends real-time heterogeneous data and asset-related information to the message broker where these information are stored for further applications such as simulations and solution generation processes. Finally, the engagement API exchanges information through a web/mobile-based interface between users and the DT system.

Once the site parameters are derived from sensory data such as point-cloud mapping or manual evaluation by an experienced project manager, the building information model (BIM) can be created for proper resource allocation for materials, equipment, and manpower. After the procurement and installation phase, sensors embedded in the physical product can be translated into the simulation program to achieve real-time monitoring and control with strict adherence to the various protocols. During the course of construction, the DT established is capable of remote monitoring and controlling the slewing unit of the tower crane. By leveraging the existing

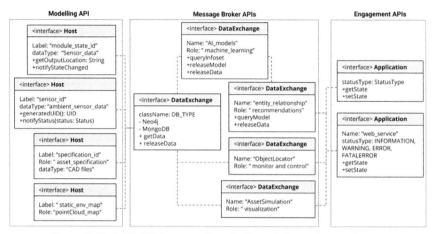

Figure 7.11 A DT API framework to facilitate information exchange.

data representation model and evaluation model, predictive maintenance and timely change of its components can be made rather readily based on the real-time sensing data. Meanwhile, having a precise model would also enable operators to identify fault locations for maintenance work in the future. With embedded sensors monitoring not just the crane but also environmental factors such as visibility and wind speed, safety protocols such as safety distances can be activated with emergency protocols to halt lifting work in the event of unsuitable conditions. Therefore, this system ensures a speedier delivery and enforces safety standards effectively.

7.5.2 Key advantages enabled via in-context solution design

In most construction projects, tower cranes are considered as bottleneck assets in typical construction projects due to their high cost of rental/ownership while many manpower resources are required to facilitate lifting operations. Apart from these factors, utilization of crane family assets usually begins way before the setup commencement, commonly known as the design stage. In this stage, product family design performs a significant role in reducing potential costs to fulfill an extensive range of construction requirements. Firstly, potential tower crane models must be configured via compatible product family modules with consideration for budget constraints and stakeholder expectations. Next, the proposed sites for tower crane installation must be evaluated to ensure maximum asset efficiency and minimize workflow interferences. When dealing with tower crane selection and planning processes, project managers usually employ the services of third-party experts who would manually recommend a suitable tower crane based on experience. The lengthy process could stretch to weeks as an approved surveyor would have to physically view and understand the scope of the construction project. This tedious and costly approach is inefficient and does not necessarily guarantee best possible outcomes due to the possibility of human error and changing circumstances.

Besides resolving design-related obstacles, context-aware DT systems can also assist in the tower crane usage phase to increase workflow efficiency and manage disruptions. As tower crane operations require a large team of safety and coordination specialists to ensure that relevant guidelines and regulations are met, this high cost of upkeep can be reduced by utilizing DT technologies to automate lifting processes through the use of real-time monitoring and control capabilities. Leveraging on the virtual replica of the construction site and predefined safety boundaries, remote operators can better visualize, and control crane assets based on the extensive 3D overview of both the construction site and tower crane. This offers a clearer view of current lifting operations and allows the operator to better supervise and control asset workflows. Moreover, embedded sensors and computational tools facilitate prognostics and health maintenance (PHM) services by performing inspection and condition monitoring tasks while the DT-driven decision-support system renders assistance ranging from automated path planning and disruption management throughout the entire project (Cai, Cai, Chandrasekaran, & Zheng, 2016). Fig. 7.12 recaps the key components to establish a DT-driven smart crane system along with their capabilities.

Although many existing DT studies have proposed innovative approaches to tackle related problems, these solutions mostly target specific PLM aspects and do not factor in external disruptions which typically occur in dynamic environments. As such, a generic DT framework incorporating ambient information is required to tackle these challenges and satisfy stakeholder expectations in a holistic manner. This DT-enhanced approach must be able to expedite both product family asset planning and on-site usage phases, with considerations to enhance asset life cycle throughout the project timeframe plus an added emphasis on reusability for similar projects in future. It is important to note that while this case study utilizes a tower crane, just like any other high-valued manpower intensive equipment, these assets are often a source of bottlenecks within workflows due to their limited availability. In the manufacturing realm for instance, CNC

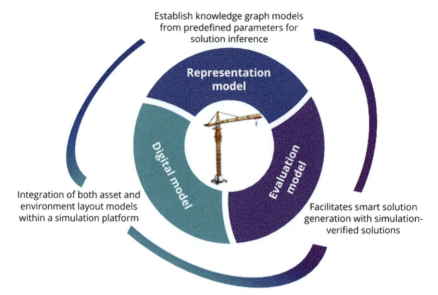

Figure 7.12 A breakdown of the trimodel-based tower crane DT system.

machines and industrial 3D printers are common examples in which context-aware DT-driven approaches could be applied.

7.6 Chapter summary

As more enterprises seek to leverage smart enabling technologies for expediting product-service design and workflow optimization processes to achieve mass customization paradigms, this study presents a context-aware DT system designed to support product family approaches with considerations to extended product life cycles. To provide stakeholders with a context-aware testbed for product evaluation, the DT system features the integration of various technologies such as IoT, knowledge graphs, and other cyber-physical aspects based on a trimodel-based approach. Fundamentally, the system was fashioned to facilitate in-context virtual prototyping and optimization of complex product family assets in the design and usage stage, respectively. Benchmark and interaction mechanisms were also presented, using knowledge graphs to facilitate family re-/design and re-/configuration solution generation. Following that, a case study showcasing a prevailing industry inefficiency pain point was resolved, demonstrating the capabilities of the context-aware DT system.

To sum up the main contributions of this study, firstly, a generic trimodel-based approach was introduced featuring the integration of the digital, representation, and computation models to enable context awareness in DT systems. Secondly, the benchmark and interaction mechanisms provided an ambient-based testbed for efficient product family prototyping and planning. Lastly, the use of a knowledge graph-enabled family redesign and reconfiguration process was presented along with a smart tower crane case study to evaluate its effectiveness. Potential future works may include: (1) development of AI methodologies and algorithms to enhance smart solution generation in context aware systems; and (2) further verification of utilizing DT-enhanced approach beyond engineering product re-/configuration along the PLM in a holistic and sustainable manner.

References

Baheti, R., & Gill, H. (2011). Cyber-physical systems. *The Impact of Control Technology, 12*, 161−166.

Biancolini, M. E., & Cella, U. (2018). Radial basis functions update of digital models on actual manufactured shapes. *Journal of Computational and Nonlinear Dynamics, 14*(2), 021013. https://doi.org/10.1115/1.4041680.

Browning, T. R. (2016). Design structure matrix extensions and innovations: A survey and new opportunities. *IEEE Transactions on Engineering Management, 63*(1), 27−52. https://doi.org/10.1109/TEM.2015.2491283.

Cai, P., Cai, Y., Chandrasekaran, I., & Zheng, J. (2016). Parallel genetic algorithm based automatic path planning for crane lifting in complex environments. *Automation in Construction, 62*, 133−147. https://doi.org/10.1016/j.autcon.2015.09.007.

Cerrone, A., Hochhalter, J., Heber, G., & Ingraffea, A. (2014). On the effects of modeling as-manufactured geometry: Toward digital twin. *International Journal of Aerospace Engineering, 2014*.

Dahmus, J. B., Gonzalez-Zugasti, J. P., & Otto, K. N. (2001). Modular product architecture. *Design Studies, 22*, 409−424.

Damjanovic-Behrendt, V., & Behrendt, W. (2019). An open source approach to the design and implementation of digital twins for smart manufacturing. *International Journal of Computer Integrated Manufacturing, 00*(00), 1−19. https://doi.org/10.1080/0951192X.2019.1599436.

David, J., Lobov, A., & Lanz, M. (2019). Attaining learning objectives by ontological reasoning using digital twins. *Procedia Manufacturing, 31*, 349−355. https://doi.org/10.1016/j.promfg.2019.03.055.

D'Souza, B., & Simpson, T. W. (2003). A genetic algorithm based method for product family design optimization. *Engineering Optimization, 35*(1), 1−18. https://doi.org/10.1080/0305215031000069663.

Grieves, M. (2011). *Virtually perfect: Driving innovative and lean products through product lifecycle management*. Space Coast Press.

Grieves, M. (2014). *Digital twin: Manufacturing excellence through virtual factory replication*. Retrieved from: http://innovate.fit.edu/plm/documents/doc_mgr/912/1411.0_Digital_Twin_White_Paper_Dr_Grieves.pdf.

Hochhalter, J., Leser, W. P., Newman, J. A., Gupta, V. K., Yamakov, V., Cornell, S. R., ... Heber, G. (2014). *Coupling damage-sensing particles to the digitial twin concept*.

Jiao, J., Simpson, T. W., & Siddique, Z. (2007). Product family design and platform-based product development: A state-of-the-art review. *Journal of Intelligent Manufacturing, 18*(1), 5–29. https://doi.org/10.1007/s10845-007-0003-2.

Lee, E. A. (2008). Cyber physical systems: Design challenges. *2008 11th IEEE international symposium on object and component-oriented real-time distributed computing (ISORC)*. https://doi.org/10.1109/ISORC.2008.25.

Li, J. H., Lin, L., Chen, D. P., & Ma, L. Y. (2015). An empirical study of servitization paradox in China. *The Journal of High Technology Management Research, 26*(1), 66–76. https://doi.org/10.1016/j.hitech.2015.04.007.

Lim, K. Y. H., Zheng, P., & Chen, C. (November 2019). A state-of-the-art survey of digital twin: Techniques, engineering product lifecycle management and business innovation perspectives. *Journal of Intelligent Manufacturing*. https://doi.org/10.1007/s10845-019-01512-w.

Lim, K. Y. H., Zheng, P., Chen, C., & Huang, L. (August 2020). A digital twin-enhanced system for engineering product family design and optimization. *Journal of Manufacturing Systems, 57*, 82–93. https://doi.org/10.1016/j.jmsy.2020.08.011.

Ma, J., & Kim, H. M. (2016). Product family architecture design with predictive, data-driven product family design method. *Research in Engineering Design, 27*(1), 5–21. https://doi.org/10.1007/s00163-015-0201-4.

Moghaddam, M., Cadavid, M. N., Kenley, C. R., & Deshmukh, A. V. (September 2018). Reference architectures for smart manufacturing: A critical review. *Journal of Manufacturing Systems, 49*, 215–225. https://doi.org/10.1016/j.jmsy.2018.10.006.

Qiao, L., Efatmaneshnik, M., Ryan, M., & Shoval, S. (2017). Product modular analysis with design structure matrix using a hybrid approach based on MDS and clustering. *Journal of Engineering Design, 28*(6), 433–456. https://doi.org/10.1080/09544828.2017.1325858.

Ranjan, K. R., & Read, S. (2016). Value co-creation: Concept and measurement. *Journal of the Academy of Marketing Science, 44*(3), 290–315. https://doi.org/10.1007/s11747-014-0397-2.

Rowley, J. (2007). The wisdom hierarchy: Representations of the DIKW hierarchy. *Journal of Information Science, 33*(2), 163–180. https://doi.org/10.1177/0165551506070706.

Savarino, P., Abramovici, M., Göbel, J. C., & Gebus, P. (2018). Design for reconfiguration as fundamental aspect of smart products. *Procedia CIRP, 70*, 374–379. https://doi.org/10.1016/J.PROCIR.2018.01.007.

Schleich, B., Anwer, N., Mathieu, L., & Wartzack, S. (2017). Shaping the digital twin for design and production engineering. *CIRP Annals - Manufacturing Technology, 66*(1), 141–144. https://doi.org/10.1016/j.cirp.2017.04.040.

Schluse, M., Priggemeyer, M., Atorf, L., & Rossmann, J. (2018). Experimentable digital twins-streamlining simulation-based systems engineering for industry 4.0. *IEEE Transactions on Industrial Informatics, 14*(4), 1722–1731. https://doi.org/10.1109/TII.2018.2804917.

Simpson, T. W. (2004). Product platform design and customization: Status and promise. *Artificial Intelligence for Engineering Design, Analysis and Manufacturing, 18*(1), 3–20. https://doi.org/10.1017/S0890060404040028.

Simpson, T. W., Jiao, J. R., Siddique, Z., & Hölttä-Otto, K. (2014). *Advances in product family and product platform design: Methods & applications*. https://doi.org/10.1007/978-1-4614-7937-6.

Takenaka, T., Yamamoto, Y., Fukuda, K., Kimura, A., & Ueda, K. (2016). Enhancing products and services using smart appliance networks. *CIRP Annals - Manufacturing Technology, 65*(1), 397–400. https://doi.org/10.1016/j.cirp.2016.04.062.

Tao, F., Cheng, J., Qi, Q., Zhang, M., Zhang, H., & Sui, F. (2017). Digital twin-driven product design, manufacturing and service with big data. *International Journal of Advanced Manufacturing Technology*. https://doi.org/10.1007/s00170-017-0233-1.

Tseng, M. M., Jiao, R. J., & Wang, C. (2010). Design for mass personalization. *CIRP Annals - Manufacturing Technology, 59*(1), 175−178. https://doi.org/10.1016/j.cirp.2010.03.097.

Tuegel, E. (2012). The airframe digital twin: Some challenges to realization. In *53rd AIAA/ASME/ASCE/AHS/ASC structures, structural dynamics and materials conference 20th AIAA/ASME/AHS adaptive structures conference 14th AIAA* (p. 1812).

Wang, Z., Chen, C. H., Zheng, P., Li, X., & Khoo, L. P. (2019). A graph-based context-aware requirement elicitation approach in smart product-service systems. *International Journal of Production Research, 0*(0), 1−17. https://doi.org/10.1080/00207543.2019.1702227.

Wang, D., Du, G., Jiao, R. J., Wu, R., Yu, J., & Yang, D. (2016). A Stackelberg game theoretic model for optimizing product family architecting with supply chain consideration. *International Journal of Production Economics, 172*, 1−18. https://doi.org/10.1016/j.ijpe.2015.11.001.

Wei, W., Tian, Z., Peng, C., Liu, A., & Zhang, Z. (2019). Product family flexibility design method based on hybrid adaptive ant colony algorithm. *Soft Computing, 23*(20), 10509−10520. https://doi.org/10.1007/s00500-018-3622-y.

Wu, Z., Kwong, C. K., Lee, C. K. M., & Tang, J. (2016). Joint decision of product configuration and remanufacturing for product family design. *International Journal of Production Research, 54*(15), 4689−4702. https://doi.org/10.1080/00207543.2015.1109154.

Yao, X., Moon, S. K., & Bi, G. (April 2016). A cost-driven design methodology for additive manufactured variable platforms in product families. *Journal of Mechanical Design, 138*, 1−12. https://doi.org/10.1115/1.4032504.

Zhang, J., Xue, D., & Gu, P. (2015). Adaptable design of open architecture products with robust performance. *Journal of Engineering Design, 26*(1−3), 1−23. https://doi.org/10.1080/09544828.2015.1012055.

Zheng, P., Chen, C., & Shang, S. (January 2019). Towards an automatic engineering change management in smart product- service systems − a DSM-based learning approach. *Advanced Engineering Informatics, 39*, 203−213. https://doi.org/10.1016/j.aei.2019.01.002.

Zheng, P., & Lim, K. Y. H. (2020). Product family design and optimization: A digital twin-enhanced approach. In *Procedia CIRP* (Vol. 93, pp. 246−250). Elsevier B.V. https://doi.org/10.1016/j.procir.2018.02.026.

Zheng, P., Lin, T.-J., Chen, C.-H., & Xu, X. (2018). A systematic design approach for service innovation of smart product-service systems. *Journal of Cleaner Production*. https://doi.org/10.1016/j.jclepro.2018.08.101.

Zheng, P., & Shankar, A. (February 2020). Full length article A generic tri-model-based approach for product-level digital twin development in a smart manufacturing environment. *Robotics and Computer-Integrated Manufacturing, 64*, 101958. https://doi.org/10.1016/j.rcim.2020.101958.

Zheng, P., Xu, X., & Chen, C.-H. (2018). A data-driven cyber-physical approach for personalised smart, connected product co-development in a cloud-based environment. *Journal of Intelligent Manufacturing*, 1−16.

Zheng, P., Xu, X., Yu, S., & Liu, C. (2017). Personalized product configuration framework in an adaptable open architecture product platform. *Journal of Manufacturing Systems, 43*, 422−435. https://doi.org/10.1016/j.jmsy.2017.03.010.

CHAPTER 8

Engineering lifecycle implementations of smart product-service system

Contents

8.1 Design stage	183
8.2 Manufacturing stage	185
8.3 Distribution stage	187
8.4 Usage stage	189
8.5 End-of-life stage	191
8.6 Application scenarios	193
8.7 Chapter summary	196
References	196

So far, the authors have already discussed how to develop the Smart PSS by addressing the three key questions holistically in the previous chapters. In this chapter, we will further explore the major implementations of Smart PSS to discover how it generates new values and influences the engineering lifecycle management. Considering the fact that advanced digital technologies enable the innovations of Smart PSS, this chapter mainly focus on the technical task classification and application scenarios of Smart PSS rather than the organizational changes from business perspectives, by summarizing the existing cases along different engineering lifecycle stages, including *design stage, manufacturing stage, distribution/logistic stage, usage stage*, and *end-of-life stage* (Stark, 2015, pp. 1–29; Zheng, Lin, Chen, & Xu, 2018b), as shown in Fig. 8.1.

The *digitalization dimension* (i.e., the vertical axis in blue) schematically represents that the Smart PSS will be developed based on a cloud-based and cyber-physical system-based platform. Particularly, the digital twin of the objects/entities of Smart PSS (i.e., the double-headed arrows), as the core of the Smart PSS development platform, is supposed to be constructed between the objects/entities in the physical world and the virtual models in the cyber world. The analytics and the decisions made in the platform should count on massive product-sensed data and user-generated data. The data from individual PSBs which is local, small-scale, and short-term will be

---

Smart Product-Service Systems
ISBN 978-0-323-85247-0
https://doi.org/10.1016/B978-0-323-85247-0.00008-6

© 2021 Elsevier Inc.
All rights reserved.

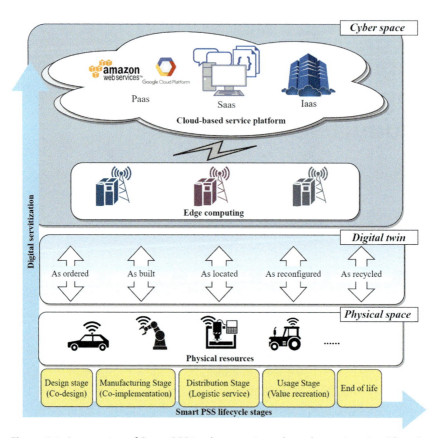

Figure 8.1 An overview of Smart PSS implementations along the engineering lifecycle.

processed by the edge computing solutions, while the data from the PSBs network/information platform which is distributed, large-scale, and long-term will be stored, handled, and retrieved in the cloud computing solutions. Accordingly, the service providers can analyze the total data and make further decisions back on the end devices for better service delivery.

The *lifecycle dimension* (i.e., the horizontal axis in blue) displays the implementation of Smart PSS along the engineering lifecycle stages. The constructed digital twins of the PSS-related objects/entities enable the design activities as ordered, the manufacturing processes as built, the distribution services as located and planned, the extended usage durations as maintained/reconfigured, and the end-of-life options as recycled. Each stage can be prolonged with user participations actively or unintentionally, such as the automatic concept evaluation and engineering change management, or the predictive maintenance services.

Based on the aforementioned framework, the selected research works are obtained from Web of Science (WoS), based on the following query sentence: "Topic = (PSS OR "PRODUCT-SERVICE SYSTEMS" OR SERVITIZATION) AND Topic = (SMART OR DIGITAL); Time Span: 2014 − 2020; Language: English". WoS was chosen as the reliable database because of its broad coverage on the major peer-reviewed articles in academia. The time span is defined by the first coin of Smart PSS to date. After initial screening, 88 relevant publications from WoS are selected (accessed on November 30, 2020), and they are further categorized into the following engineering lifecycle stages.

8.1 Design stage

Smart PSS facilitates three types of activities in the design stage, namely *requirement management, concept generation,* and *concept evaluation,* as listed in Table 8.1. Many attempts in labs and even in industries have already shown its potentials with the supports of many advanced digitalized technologies, such as digital twin (Schleich, Anwer, Mathieu, & Wartzack, 2017; Tao et al., 2017), augmented reality (AR) (Gupta et al., 2018), 3D printing (Oztemel & Gursev, 2020), computer simulation (Cheng, Zhang, Tao, & Juang, 2020) cyber-physical systems (Tao & Qi, 2019; Wiesner et al., 2017), and online customization toolkit (Luchs, Swan, & Creusen, 2016).

Major literature on *requirement management* in Smart PSS study the requirement capture and evaluation process, in which the user-generated content and product-sensed data in the usage phase are also applied (Kuhlenkötter et al., 2017; Tao & Qi, 2019; Wang et al., 2018). In this way, the designers can precisely know better what the users want from the user-generated data instead of their empirical knowledge (Tao & Qi, 2019). Wang et al. (2018) used product specification data and user comments in the usage phase for requirement elicitation based on a graph-based model. Similarly, Lin Liu, Zhou, Liu, and Cao (2017) applied user behavioral data with informatics-based analytic tools for requirement elicitation. Digitalization technologies can be seen in requirement management as well. For instance, Gupta et al. (2018) exploited the AR technique to inspect the partially or fully assembled aircraft modules for the aim of requirement gathering. Verdugo Cedeño et al. (2018) proposed a customer-oriented approach for requirement identification and e-service development based on Internet of Things (IoT) platforms.

184   Smart Product-Service Systems

Table 8.1 Summary of adoptions in design stage.

Perspective	Description	References
Design tasks in Smart PSS	Requirement management	Chang, Gu, Li, and Jiang (2019), Gupta et al. (2018), Lee, Chen, and Lee (2020), Liu and Ming (2019a), Lee and Kao (2014), Liu, Ming, Qiu, Qu, and Zhang (2020), Song (2017), Takenaka, Yamamoto, Fukuda, Kimura, and Ueda (2016), Tao and Qi (2019), Verdugo Cedeño, Papinniemi, Hannola, and Donoghue (2018), Wang, Zheng, Chen, and Khoo (2018), Wiesner et al. (2017), Zheng, Ming, Wang, Yin, and Zhang (2017), Zhang, Qin, Li, Zou, and Ding (2020)
	Concept generation	Chang et al. (2019), Corradi et al. (2019), Lee, Chen, and Trappey (2019), Li et al. (2020), Lim, Kim, Heo, and Kim (2015), Poeppelbuss and Durst (2019), Schleich et al. (2017), Tao et al. (2017), Watanabe, Okuma, and Takenaka (2020), Zheng, Lin, Chen, and Xu (2018a), Zheng et al. (2018b), Zhang et al. (2020)
	Concept evaluation	Chang et al. (2019), Chen and Ming (2020), Chen and Ming (2020), Cong, Chen, and Zheng (2020), Liu, Song, and Han (2020a), Liu, Lin, et al. (2019), Takenaka et al. (2016), Tao et al. (2017), Zheng, Chen, and Shang (2019)

To achieve creative *concept generation*, an increasing number of design toolkits have been attempted to enhance design smartness. Facing the difficulty of professional consultations in traditional PSS, Li et al. (2020) adopted the knowledge graph (KG) and concept-knowledge (C-K) model for evolutionary design generation. To integrate the physical properties of products and use behavior into the cyberspace, Tao et al. (2017) displayed an example of bicycle design refinement based on a digital twin. Furthermore, Zheng et al. (2018b) illustrated a case study of bicycle that digital twin serves as the integration between physical product and the CAD model, AR assists the monitoring of user's riding status, and cloud-based platform achieves the prospective design improvement. To represent the concepts, reference models are further fostered as well. For example, Schleich et al. (2017) proposed a comprehensive reference model serving as a digital twin of the physical product in design for concept representation and implementation. Corradi et al. (2019) also enabled the novel service design for different stakeholders based on the reference architecture model industry 4.0 (RAMI 4.0).

The *concept evaluation* in Smart PSS can be largely strengthened by artificial intelligence technologies, including natural language processing (NLP) techniques and machine-learning algorithms. On one hand, NLP techniques relieve the linguistic ambiguousness of experts and users which always appeared in the conventional PSS. Chen and Ming (2020) formalized the module selection tasks as a multicriteria decision-making problem and considered the experts' linguistic ambiguousness and their preference randomness in the real world. Similarly, Lingdi Liu, Song et al. (2020) also considered the randomness vagueness experts' judgments and transformed the Smart PSS designs' sustainability assessment as a multicriterion decision-making problem. On the other hand, machine-learning approaches, e.g., support vector machine, have already been attempted by Long, Wang, Shen, Wu, and Jiang (2013) for solution selection. Lim et al. (2015) displayed a case study of driving safety enhancement service based on informatics-based analytic tools. Other examples can also be seen in the study of Zheng, Chen, et al. (2019) that an occurrence-based DSM approach was proposed for engineering change management based on the historical in-context engineering change logs and Louvain algorithm.

8.2 Manufacturing stage

In the manufacturing stage, the idea of Smart PSS to offer user-required functionalities with lower environmental effects in a more sustainable manner is concreted as the concept "Manufacturing-as-a-Service"; meanwhile, it is organized by an IT-driven service-oriented smart manufacturing (SoSM) framework in the manufacturing stage proposed by Tao and Qi (2019). The SoSM framework aims to realize smart manufacturing in which the interoperations between devices, the integration of knowledge, and data fusion are required in the cyber-physical world of manufacturing.

To achieve it, *IoT, big data, cloud computing,* and *cyber-physical systems* (*CPS*) are deployed in the proposed SoSM framework. By applying IoT, the ubiquitous connection among devices can be achieved, thus enabling the heterogeneous data collection from nearly each manufacturing process.

Those data can then be converged and continuously stored in the clouds. With big data analytics technology, the raw data can be cleaned and analyzed to extract valuable information and knowledge. By analyzing the information and learning from the knowledge, the manufacturing enterprises have the opportunities to better understand the manufacturing performance, such as yield, product quality, and the flexibility of the production lines to the customer demands changes. Those capabilities can be encapsulated into the value-added services to the manufacturing companies. By precepting, receiving, and handling the large-scale data from the physical devices and humans in the cyber world, the CPS connects the cyber space and the physical space via the connected data. The analyzed results and their corresponding operations can be fed back to control the physical space.

Those technologies largely enable the Smart PSS adaptions in manufacturing (Herterich, Uebernickel, & Brenner, 2015; Mikusz, 2014), particularly in *smart production* and *smart inspection*, as shown in Table 8.2.

Smart production mainly contains resource management, production plan, and production process control (Tao et al., 2017). Based on the collected literature, digital twin was emphasized by many researchers in different manufacturing tasks. For instance, Uhlemann et al. (2017) claimed that the digital twin can support the simulation-based production process optimization. Moreover, Tao et al. demonstrated a resource management plan based on a digital twin of the shop floor in (Tao & Qi, 2019), in which the digital twin performs as the basic executor of the product resources organization. They also manifested the production process monitoring and

Table 8.2 Summary of adoptions in manufacturing stage.

Task category	Specification	References
Smart production	Resource management	Klein, Biehl, and Friedli (2018), Tao et al. (2017)
	Production planning	Filho, Liao, Loures, and Canciglieri (2017), Liu, Zhang, et al. (2019), Schleich et al. (2017), Tao and Qi (2019), Tao et al. (2017), Uhlemann, Lehmann, and Steinhilper (2017)
	Production process control	Aheleroff et al. (2020), Brad and Murar (2015), Tao and Qi (2019), Tao et al. (2017)
Smart inspection	Geometry assurance	Schleich et al. (2017), Söderberg, Wärmefjord, Carlson, and Lindkvist (2017)

control via digital twin in (Tao et al., 2017), where the shopfloor adjustment can be generated as digitalized services by the Shop Floor Service system based on various types of data.

Smart inspection intends to minimize the geometrical variations of the final product. Toward the Smart PSS adoptions in this task, Söderberg et al. (2017) constructed a digital twin of sheet metal assembly for real-time geometry assurance, highlighting the necessity of functionality and data models and displaying the role of digital twin during the design and manufacturing stage.

8.3 Distribution stage

In the distribution stage, more and more companies have realized that well-organized and low-cost logistics plays a crucial role in the business success among the international competitiveness. Improving resource productivity has already been the ultimate goal of logistics for decades. Nowadays, the logistics organizations are moving toward more collaborative, intelligent, and service-oriented (Pan, Zhong, & Qu, 2019). More specifically, they are seeking the logistic services with higher agility, flexibility, sustainability, and resilience. As a consequence, the logistics organizations are evolving from in-house to outsourced logistics, from vertical to horizontal collaboration, and becoming more intelligent and interoperable (Mason, Lalwani, & Boughton, 2007; Pan, Trentesaux, Ballot, & Huang, 2019; Wilding & Juriado, 2004).

Under the trend of collaboration, intelligence, and servitization, Smart PSS concept can be adopted in many logistic services very well, as listed in Table 8.3. Its concept can affect logistic services design and implement from

Table 8.3 Summary of adoptions in distribution stage.

Task category	Specification	References
Smart Logistic	PI–Container management	Pan, Zhong, et al. (2019)
	PI–hubs management	Huang, Li, and Thürer (2019), Pan, Zhong, et al. (2019)
	PI information management	Pan, Zhong, et al. (2019), Shao, Xu, and Li (2019), Vazquez-Martinez, Gonzalez-Compean, Sosa-Sosa, Morales-Sandoval, and Perez (2018), Zhang, Liu, Liu, and Li (2016)

two perspectives: *servitization* and *digitalization*. For the prior one, PSS itself is a service-oriented business model who aims at generating values/profits from functionalities rather than the ownership. The PSS business model recommends the logistic companies to offer asset-free flow orchestrators for higher collaborative mutual supply chain (Pan, Zhong, et al., 2019). For the latter one, by making the entities (e.g., the containers) as smart, connected products, and integrating with specialized analytic tools, Smart PSS enables automatic data collection and analysis for decision-makings and even the self-adaptive actions. Intelligence will be more effective and efficient for the resource management in the interoperable supply chain. More importantly, the digitalization of Smart PSS also offers the logistics organizations a radically different resource management means with comprehensive integration among multiple parties and continuous monitoring of the products/services, thus improving sustainability and productivity (Bechtsis, Tsolakis, Vlachos, & Iakovou, 2017).

The Smart PSS adoptions can be seen in many logistic-related studies. Pan, Zhong, et al. (2019) proposed a three-tier framework to discuss how to deploy Smart PSS in logistics from three levels, using physical internet (PI) as the example. At the most specific level, PI-containers were designed as smart, connected products with intelligence, which are capable of collecting information (first-level intelligence), notifying problems (second-level intelligence), and making decisions (third-level intelligence). Those intelligent services with real-time decentralized and distributed decision-makings are the add-on values for customers with better service quality although probably with higher costs.

At the PI-hubs level, unlike the long-term contracts among logistics participants in the traditional supply chain, pay-per-use model is recommended to the asset-free service providers (e.g., 4 PL) that a charter containing several business protocols will be respected by all parties, making the same services acceptable and executable by any companies anywhere. Even smarter contract for supply chain was proposed by Huang et al. (2019) that blockchain can synchronize all the transaction data when the smart contracts are allowed.

Finally, at the level of PI's information management, lots of information tools, such as databases, platforms, intelligent algorithms, blockchains, and other services, are applied and can be regarded as the entities constituting the intelligent services. Two issues were summarized in this level, namely object monitoring and data sharing. Object monitoring refers to the task of tracking objects from one party to another. Its faces the challenges of data unavailability and violations of confidentiality. Vazquez-Martinez et al. (2018) presented a novel distribution model called CloudChain with seamless information flow between customers, partners and organizations to support the information integration during the product lifecycle. Also aiming to achieve more transparent real-time information trajectory, Zhang et al. (2016) designed a smart box-enabled PSS platform based on cloud computing which has the functions of order-box matching, order clustering, and real-time optimization for logistic orders to achieve least unused containers. Instead of the traceability of products, Shao et al. (2019) considered the traceability of the planned vehicle routes and proposed a digital twin-enabled route optimization system by updating user requirements and the transport network data from IoT devices during delivery process.

As a result, a global scope of logistic, more transparent information trajectory and more sustainable services can be achieved with the adoption of PI (Zhang et al., 2016), cloud computing (Vazquez-Martinez et al., 2018), blockchain (Huang et al., 2019), and even other advanced technologies into the Smart PSS model.

8.4 Usage stage

Numerous Smart PSS adoptions in different stages of the product lifecycle are driven by the user-generated data or product-sensed data from the usage stage. The usage stage of the smart products itself, at the same time, is also dramatically changed under the Smart PSS paradigm. Lots of studies discussed the Smart PSS adoptions in usage phase, they can be further classified into two clusters, including *smart operation/maintenance* and *smart reconfiguration*, as listed in Table 8.4.

Smart operation/maintenance does not radically change the original designs but monitor, assess, and maintain the current designs. It primarily comprises two types of specific tasks, i.e., performance assessment and maintenance/diagnosis. Performance assessment mainly concerns about the analysis of current PSBs, e.g., the riding speed and riding time of a bicycle (Tao et al., 2017); while the maintenance/diagnosis emphasizes locating the problems and responding to them which requires PSB's actions (Alcayaga et al., 2019), e.g., the prognostic and health management (PHM) of a machine (Lee et al., 2015). In Smart PSS paradigm, with the supports of many advanced enabling technologies, even more intelligent (e.g., context-aware, automatic, or customized) performance assessment or maintenance services can be armed to the PSBs. For example, a digital twin-based remote monitoring and

190 Smart Product-Service Systems

Table 8.4 Summary of adoptions in usage stage.

Task category	Specification	References
Smart operation/ maintenance	Performance assessment	Lee, Ardakani, Yang, and Bagheri (2015), Lim et al. (2015), Lu, Lai, and Liu (2019), Tao et al. (2017), Zheng et al. (2018b)
	Maintenance, diagnosis	Alcayaga, Hansen, and Wiener (2019), Grubic and Jennions (2018), Lee, Kao, and Yang (2014), Lee et al. (2015), Liu and Ming (2019b), Maleki et al. (2018), Mourtzis, Zogopoulos, and Vlachou (2017), Mourtzis, Vlachou, and Zogopoulos (2017), Karmann, Rudolf, Sommer, Schuh, and Pitsch (2014), Tao et al. (2017), Tao, Zhang, Liu, and Nee (2018)
Smart reconfiguration	Product reconfiguration management	Abramovici, Göbel, and Dang (2016), Abramovici, Göbel, and Savarino (2017), Gembarski (2020), Guan, Christensen, Vyatkin, Dai, and Dubinin (2016)

maintenance platform for complex products was proposed by Tao et al. (2017), aiming to reduce the effects of the assumed similitude between the underlying statistical situations and the practical complex product's operation situations. They further deepened the digital twin-based PHM platform in (Tao et al., 2018) with a novel PHM method and validated it with a wind turbine example, for the aim of improving accuracy of PHM. Lee et al. (2015) designed a CPS-based platform for complex products (using a CNC as a case) with the functions of usage data analysis, health condition assessment, and predicted maintenance. Besides, AR was also attempted for remote monitoring and maintenance by Mourtzis, Zogopoulos, et al. (2017), Mourtzis, Vlachou, et al. (2017). They deployed a cloud-based platform with the support of AR technology that can generate AR scenes in an automated way for the on-spot technician for diagnosis. Furthermore, facing the multifarious specifications of the complex products and their related services, ontology technology was also applied for the monitoring and maintenance as well to extract and reuse the domain knowledge. Maleki et al. (2018) defined a sensor ontology as the backbone of the Smart PSS, the ontology dedicates to managing the whole life cycle of the designed PSS.

Smart reconfiguration is described as the modifications of an existing smart product to meet new requirements (Abramovici et al., 2017). Owing to the dynamic reconfiguration capability and high degree of personalization feature of the smart products and the digitalized services, reconfiguration in Smart PSS are strengthened from two perspectives: in-context change management and digitalized service change (Zheng, Chen, et al., 2019). For the prior one, the conventional reconfiguration methods mainly spend a long time on the marketing analysis tools (e.g., questionnaire, focus group) or the user feedbacks collection during the usage phase, making the modification on PSBs delayed from the user requirement changes. But in a smart and connected environment, real-time product-sensed data and user feedbacks are both accessible for reconfiguring PSB's functions (Zheng, Chen, et al., 2019). For the latter one, not only the physical products can be reconfigurable, but also the digitalized services serving as the bundle of software and hardware are supposed to be reconfigured during the usage stage as well. Hence, the virtual model of PSBs should be taken into consideration. These progresses were also highlighted by Abramovici et al. (2016), (2017) and Gembarski (2020). Particularly, Abramovici et al. (2016) proposed a conceptual framework to represent how to continuously conduct smart product reconfiguration based on a digital product twin containing the virtual product model and the user feedbacks from the physical product.

8.5 End-of-life stage

After using PSBs over a period of time, end-of-life options should be considered when some physical components approach the end of lifetime that PSBs cannot continuously and smoothly run. Considering the sustainability with less material and energy consumption and prolonged product usable lifetime, several end-of-life options are discussed, including recycle, reuse, and remanufacturing. With the support of advanced technologies meanwhile under the context of Smart PSS, the end-of-life strategies will become more interconnected and more intelligent as *smart reuse*, *smart recycling*, and *smart remanufacturing*, as listed in Table 8.5. Few study discussed the Smart PSS adoptions in product's end-of-life stage, but Alcayaga et al. (2019) discussed in depth a novel framework called "smart-circular" by integrating Smart PSS and circular economy into an overall consideration.

192 Smart Product-Service Systems

Table 8.5 Summary of adoptions in end-of-life stage.

Task category	Specification	References
Smart reuse	Product reuse	Alcayaga et al. (2019)
Smart recycling	Product recycling	Alcayaga et al. (2019)
Smart remanufacturing	Product remanufacturing	Alcayaga et al. (2019), Kerin and Pham (2020)

Smart reuse can be generally regarded as the nondisruptive process that allows a second, third, or even more usage cycle of products without changing its physical design. Seldom literature discussed the reuse strategies with intelligent and digitalization capability (Alcayaga et al., 2019). Alcayaga et al. (2019) offered three smart reuse strategies. Firstly, by monitoring the products' lifetime during usage stage, the product can be directly recycled and reused. This suggestion is more suitable for the unpowered or low-tech products, e.g., clothes or packages (Condea, Thiesse, & Fleisch, 2010). Secondly, using IoT technology and the identification labels (e.g., RFID tags or QR codes) can (1) reduce the emissions and pollution during the process of recycled products classification, and (2) relocate products to the nearby users with lower cost (Jun, Shin, Kim, Kiritsis, & Xirouchakis, 2009). Thirdly, an integrated information platform with more transparent data transmission will further help the associated parties to evaluate how much value a product has for reuse. Obviously, the second and third smart reuse strategies accord with the digitalization trend in the Smart PSS context.

Smart recycling during end-of-life aims at collecting unused products and extracting raw materials from products to use them in new products (Alcayaga et al., 2019). In the proposed smart-circular concept combing Smart PSS and circular economy, the smart recycling tasks can be enhanced in both the service efficiency and the service quality. On the one hand, the recycling process can be even more effective and efficient by deploying IoT technologies and the accessible product lifetime management. By embedding identification labels and sensors, the smart products themselves, rather than an extra database, can offer the product composition (e.g., material information) and lifetime information, thus simplifying the process of waste product collection (Vadde, Kamarthi, Gupta, & Zeid, 2008). On the other hand, the recycling service quality can be improved as well. Easier sorting process can be enabled by early differentiation labels, making higher volume and higher purity of recycled materials recovered (Binder, Quirici, Domnitcheva, & Stäubli, 2008). At the same time, an optimized recycling collection route can be easily planned based on the smart products' composition information and the lifetime data. With new capabilities generated in Smart PSS, the performance of the recycling tasks can be improved, making the product lifecycle become a close loop.

Remanufacturing intends to alter the products into the as-new ones with at least equivalent performance. Smart PSS adoptions facilitate the *smart remanufacturing* process and improve the quality of the remanufactured products by many advanced technologies, including robotics, additive manufacturing, big data, artificial intelligence, advanced materials, and identification labels (Kerin & Pham, 2020). It is prudent to explain that those technologies are not restricted to facilitate the smart remanufacturing process.

8.6 Application scenarios

Table 8.6 summaries the typical application sectors, scenarios, and their descriptions of Smart PSS, and it is interesting to find that though most of the applications are still within the smart manufacturing and smart living sectors. However, gradually they are moving to other sectors such as smart city and smart business as well. Some points in these emerging fields can be summarized as: (1) Smart PSS provides new capabilities for online and offline service innovations of the smart appliance. The quality of the appliances' performance can be ever improved with higher efficiency and guaranteed with higher flexibility and robustness. (2) Smart PSS also have emerged in the field of smart agriculture, where by integrating the agricultural equipment and relevant information together, the field of agriculture can be a smart ecosystem. (3) Though the concept of smart city (e.g., sensing-as-a-service) has already discussed from the aspect of weather, traffic, lighting, flood, and so on, the integration of smart city and Smart PSS provision in a sustainable and customized manner still remains to be investigated. (4) Most of the smart business scenarios discusses about the future business generation. However, there is still a lack of consensus on which business innovation method has the capabilities such as generalization and robustness to fit different business features.

Table 8.6 Examples of application scenarios in Smart PSS.

Sectors	Application scenarios	Reference	Description
Smart manufacturing	Digital-twin-enabled shaft manufacturing	Tao et al. (2017)	(1) Devise plan of allocating steel bars and CNC machines; (2) generate and simulate production plan; and (3) evaluate production plan throughout the manufacturing process.
	CPS-based machine tool prognostics and health management (PHM)	Lee et al. (2015)	Holistic data analysis and health condition monitor services of machine tools based on cyber-physical systems with 5C architecture.
	Digital-twin-enabled power transformer	Tao et al. (2017)	(1) Real-time state monitoring of power transformer; (2) energy consumption of power transformer itself; and (3) output power quality prediction and analysis.
Smart living/ Smart appliance	A data-driven personalized wearable mask	Zheng et al. (2018a)	(1) Cloud-based breathing condition monitoring (online service innovation); and (2) checking mask fitness (offline service innovation) during usage.
	Smart water dispenser	Zheng, Chen, et al. (2019)	Remote water dispenser maintenance service management based on crowd sensing data.
	Cyber-physical smart, connected bicycle	Zheng et al. (2018b)	A CPS-based smart open architecture bike prototype, of which performance analysis, parametric design and predictive maintenance can be achieved.
	Smart car maintenance and insurance	Zheng, Wang, et al. (2019)	(1) Notifying local repair shops nearby based on GPS once the car accident happens; and (2) providing real-time quoting based on text mining and image processing.
	Smart remote machinery maintenance systems with Komatsu	Lee et al. (2014)	(1) Remote prognostics and monitoring and (2) remaining life prediction.

Smart agriculture	Digital services for a tractor company	Verdugo Cedeño et al. (2018)	Providing tailor-made services and solutions, for example, the personalized everyday working plan based on data analyzed on previous work shifts
Smart city	Urban planning, disaster monitoring/warning	Liu, Xiao, Zhang, and Chen (2018)	Leveraging personal smart phone for conducting crowd sensing tasks in smart city.
Smart business	Innovation strategies for a mattress company	Lee and Kao (2014)	Analyzing the current scenarios and innovative scenarios of mattress that whether they satisfy visible and invisible customer requirements and offer out-of-the-box services
	Disruptive innovation strategies for a small software-oriented company	Weiβ, Kölmel, and Bulander (2016)	Disruptive innovation of a small software-oriented company, by reorientation and creating new service-based offerings for intelligent energy management.

8.7 Chapter summary

From the above description, one can find that Smart PSS have been widely adopted in the engineering fields across many sectors with impactful outcomes. It can largely strengthen the task fulfillment along each engineering lifecycle stage empowered by the support of IT infrastructure and advanced digital technologies. To date, manufacturing is still playing a dominant role among Smart PSS application scenarios owing to its advancement, while ever increasing attentions have been brought to the primary and tertiary industry to make more revenues. With ever-maturing development of high performance, low cost SCPs, various new digitalized services will be generated and brought into almost every sector in the near future. Moreover, with more electronic components been utilized, a sustainable Smart PSS implementation process should be emphasized (e.g., e-waste management) for the aim of extended lifespan, better resource efficiency, and a closed lifecycle loop, which will be further described in Chapter 9.

References

Abramovici, M., Göbel, J. C., & Dang, H. B. (2016). Semantic data management for the development and continuous reconfiguration of smart products and systems. *CIRP Annals, 65*(1), 185−188. https://doi.org/10.1016/j.cirp.2016.04.051.

Abramovici, M., Göbel, J. C., & Savarino, P. (2017). Reconfiguration of smart products during their use phase based on virtual product twins. *CIRP Annals - Manufacturing Technology, 66*(1), 165−168. https://doi.org/10.1016/j.cirp.2017.04.042.

Aheleroff, S., Xu, X., Lu, Y., Aristizabal, M., Pablo Velásquez, J., Joa, B., & Valencia, Y. (2020). IoT-enabled smart appliances under industry 4.0: A case study. *Advanced Engineering Informatics*. https://doi.org/10.1016/j.aei.2020.101043.

Alcayaga, A., Hansen, E. G., & Wiener, M. (2019). Towards a framework of smart-circular systems: An integrative literature review. *Journal of Cleaner Production, 221*, 622−634. https://doi.org/10.1016/j.jclepro.2019.02.085.

Bechtsis, D., Tsolakis, N., Vlachos, D., & Iakovou, E. (2017). Sustainable supply chain management in the digitalisation era: The impact of Automated Guided Vehicles. *Journal of Cleaner Production*. https://doi.org/10.1016/j.jclepro.2016.10.057.

Binder, C. R., Quirici, R., Domnitcheva, S., & Stäubli, B. (2008). Smart labels for waste and resource management: An integrated assessment. *Journal of Industrial Ecology, 12*(2), 207−228.

Brad, S., & Murar, M. (2015). Employing smart units and servitization towards reconfigurability of manufacturing processes. *Procedia CIRP, 30*, 498−503. https://doi.org/10.1016/j.procir.2015.02.154.

Chang, D., Gu, Z., Li, F., & Jiang, R. (2019). A user-centric smart product-service system development approach: A case study on medication management for the elderly. *Advanced Engineering Informatics*. https://doi.org/10.1016/j.aei.2019.100979.

Cheng, J., Zhang, H., Tao, F., & Juang, C. (2020). DT-II: Digital twin enhanced Industrial Internet reference framework towards smart manufacturing. *Robotics and Computer-Integrated Manufacturing, 62*, 101881. https://doi.org/10.1016/j.rcim.2019.101881.

Chen, Z., & Ming, X. (2020). A rough—fuzzy approach integrating best—worst method and data envelopment analysis to multi-criteria selection of smart product service module. *Applied Soft Computing Journal*. https://doi.org/10.1016/j.asoc.2020.106479.

Condea, C., Thiesse, F., & Fleisch, E. (2010). Assessing the impact of RFID and sensor technologies on the returns management of time-sensitive products. *Business Process Management Journal, 16*(6), 954—971.

Cong, J., Chen, C. H., & Zheng, P. (2020). Design entropy theory: A new design methodology for smart PSS development. *Advanced Engineering Informatics*. https://doi.org/10.1016/j.aei.2020.101124.

Corradi, A., Foschini, L., Giannelli, C., Lazzarini, R., Stefanelli, C., Tortonesi, M., & Virgilli, G. (2019). Smart appliances and RAMI 4.0: Management and servitization of ice cream machines. *IEEE Transactions on Industrial Informatics, 15*(2), 1007—1016. https://doi.org/10.1109/TII.2018.2867643.

Filho, M. F., Liao, Y., Loures, E. R., & Canciglieri, O. (June 2017). Self-aware smart products: Systematic literature review, conceptual design and prototype implementation. *Procedia Manufacturing, 11*, 1471—1480. https://doi.org/10.1016/j.promfg.2017.07.278.

Gembarski, P. C. (2020). The meaning of solution space modelling and knowledge-based product configurators for smart service systems. In *Advances in intelligent systems and computing*. https://doi.org/10.1007/978-3-030-30440-9_4.

Grubic, T., & Jennions, I. (2018). Remote monitoring technology and servitised strategies — factors characterising the organisational application. *International Journal of Production Research, 56*(6), 2133—2149. https://doi.org/10.1080/00207543.2017.1332791.

Guan, X., Christensen, J. H., Vyatkin, V., Dai, W., & Dubinin, V. N. (2016). Toward self-manageable and adaptive industrial cyber-physical systems with knowledge-driven autonomic service management. *IEEE Transactions on Industrial Informatics, 13*(2), 725—736. https://doi.org/10.1109/tii.2016.2595401.

Gupta, R. K., Belkadi, F., Buergy, C., Bitte, F., Da Cunha, C., Buergin, J., ... Bernard, A. (2018). Gathering, evaluating and managing customer feedback during aircraft production. *Computers and Industrial Engineering, 115*, 559—572. https://doi.org/10.1016/j.cie.2017.12.012. December 2017.

Herterich, M. M., Uebernickel, F., & Brenner, W. (2015). The impact of cyber-physical systems on industrial services in manufacturing. *Procedia CIRP, 30*, 323—328. https://doi.org/10.1016/j.procir.2015.02.110.

Huang, J., Li, S., & Thürer, M. (2019). On the use of blockchain in industrial product service systems: A critical review and analysis. In *Procedia CIRP*. https://doi.org/10.1016/j.procir.2019.03.117.

Jun, H.-B., Shin, J.-H., Kim, Y.-S., Kiritsis, D., & Xirouchakis, P. (2009). A framework for RFID applications in product lifecycle management. *International Journal of Computer Integrated Manufacturing, 22*(7), 595—615.

Karmann, W., Rudolf, S., Sommer, M., Schuh, G., & Pitsch, M. (2014). Modular sensor platform for service-oriented cyber-physical systems in the European tool making industry. *Procedia CIRP, 17*, 374—379. https://doi.org/10.1016/j.procir.2014.01.114.

Kerin, M., & Pham, D. T. (2020). Smart remanufacturing: A review and research framework. *Journal of Manufacturing Technology Management*. https://doi.org/10.1108/JMTM-06-2019-0205.

Klein, M. M., Biehl, S. S., & Friedli, T. (2018). Barriers to smart services for manufacturing companies — an exploratory study in the capital goods industry. *Journal of Business and Industrial Marketing, 33*(6), 846—856. https://doi.org/10.1108/JBIM-10-2015-0204.

Kuhlenkötter, B., Wilkens, U., Bender, B., Abramovici, M., Süße, T., Göbel, J., ... Lenkenhoff, K. (2017). New perspectives for generating smart PSS solutions - life cycle, methodologies and transformation. *Procedia CIRP, 64*, 217−222. https://doi.org/10.1016/j.procir.2017.03.036.

Lee, J., Ardakani, H. D., Yang, S., & Bagheri, B. (2015). Industrial big data analytics and cyber-physical systems for future maintenance & service innovation. *Procedia CIRP, 38*, 3−7. https://doi.org/10.1016/j.procir.2015.08.026.

Lee, C. H., Chen, C. H., & Lee, Y. C. (2020). Customer requirement-driven design method and computer-aided design system for supporting service innovation conceptualization handling. *Advanced Engineering Informatics.* https://doi.org/10.1016/j.aei.2020.101117.

Lee, C. H., Chen, C. H., & Trappey, A. J. C. (2019). A structural service innovation approach for designing smart product service systems: Case study of smart beauty service. *Advanced Engineering Informatics.* https://doi.org/10.1016/j.aei.2019.04.006.

Lee, J., & Kao, H.-A. (2014). Dominant innovation design for smart products-service systems (PSS): Strategies and case studies. In *2014 annual SRII global conference (SRII)* (pp. 305−310). IEEE. https://doi.org/10.1109/SRII.2014.25.

Lee, J., Kao, H. A., & Yang, S. (2014). Service innovation and smart analytics for Industry 4.0 and big data environment. *Procedia CIRP, 16*, 3−8. https://doi.org/10.1016/j.procir.2014.02.001.

Li, X., Chen, C.-H., Zheng, P., Wang, Z., Jiang, Z., & Jiang, Z. (2020). A knowledge graph-aided concept−knowledge approach for evolutionary smart product−service system development. *Journal of Mechanical Design.* https://doi.org/10.1115/1.4046807.

Lim, C.-H., Kim, M.-J., Heo, J.-Y., & Kim, K.-J. (2015). Design of informatics-based services in manufacturing industries: Case studies using large vehicle-related databases. *Journal of Intelligent Manufacturing*, 1−12.

Liu, G., Lin, L., Zhou, W., Zhang, R., Yin, H., Chen, J., & Guo, H. (2019). A posture recognition method applied to smart product service. In *Procedia CIRP.* https://doi.org/10.1016/j.procir.2019.04.145.

Liu, Z., & Ming, X. (2019a). A framework with revised rough-DEMATEL to capture and evaluate requirements for smart industrial product-service system of systems. *International Journal of Production Research*, 1−19. https://doi.org/10.1080/00207543.2019.1577566, 0(0).

Liu, Z., & Ming, X. (2019b). A methodological framework with rough-entropy-ELECTRE TRI to classify failure modes for co-implementation of smart PSS. *Advanced Engineering Informatics.* https://doi.org/10.1016/j.aei.2019.100968.

Liu, Z., Ming, X., Qiu, S., Qu, Y., & Zhang, X. (2020). A framework with hybrid approach to analyse system requirements of smart PSS toward customer needs and co-creative value propositions. *Computers and Industrial Engineering.* https://doi.org/10.1016/j.cie.2019.03.040.

Liu, L., Song, W., & Han, W. (2020). How sustainable is smart PSS? An integrated evaluation approach based on rough BWM and TODIM. *Advanced Engineering Informatics.* https://doi.org/10.1016/j.aei.2020.101042.

Liu, C. H., Xiao, D. K., Zhang, Q. S., & Chen, J. R. (2018). Exploring applying smart interconnection technology to product service systems. In *Proceedings - 2018 3rd international conference on mechanical, control and computer engineering, ICMCCE* (pp. 636−639). https://doi.org/10.1109/ICMCCE.2018.00140.

Liu, B., Zhang, Y., Zhang, G., & Zheng, P. (2019). Advanced Engineering Informatics Edge-cloud orchestration driven industrial smart product-service systems solution design based on CPS and IIoT. *Advanced Engineering Informatics, 42*(April), 100984. https://doi.org/10.1016/j.aei.2019.100984.

Liu, Lin, Zhou, Q., Liu, J., & Cao, Z. (2017). Requirements cybernetics: Elicitation based on user behavioral data. *Journal of Systems and Software*. https://doi.org/10.1016/j.jss.2015.12.030.

Long, H. J., Wang, L. Y., Shen, J., Wu, M. X., & Jiang, Z. B. (2013). Product service system configuration based on support vector machine considering customer perception. *International Journal of Production Research, 51*(18), 5450–5468.

Luchs, M. G., Swan, K. S., & Creusen, M. E. H. (2016). Perspective: A review of marketing research on product design with directions for future research. *Journal of Product Innovation Management, 33*(3), 320–341. https://doi.org/10.1111/jpim.12276.

Lu, D., Lai, I., & Liu, Y. (2019). The consumer acceptance of smart product-service systems in sharing economy: The effects of perceived interactivity and particularity. *Sustainability, 11*(3), 928. https://doi.org/10.3390/su11030928.

Maleki, E., Belkadi, F., Boli, N., van der Zwaag, B. J., Alexopoulos, K., Koukas, S., … Mourtzis, D. (2018). Ontology-based framework enabling smart product-service systems: Application of sensing systems for machine health monitoring. *IEEE Internet of Things Journal, 5*(6), 4496–4505. https://doi.org/10.1109/JIOT.2018.2831279.

Mason, R., Lalwani, C., & Boughton, R. (2007). Combining vertical and horizontal collaboration for transport optimisation. *Supply Chain Management*. https://doi.org/10.1108/13598540710742509.

Mikusz, M. (2014). Towards an understanding of cyber-physical systems as industrial software-product-service systems. *Procedia CIRP, 16*, 385–389. https://doi.org/10.1016/j.procir.2014.02.025.

Mourtzis, D., Vlachou, A., & Zogopoulos, V. (2017b). Cloud-based augmented reality remote maintenance through shop-floor monitoring: A product-service system Approach. *Journal of Manufacturing Science and Engineering, 139*(6), 061011. https://doi.org/10.1115/1.4035721.

Mourtzis, D., Zogopoulos, V., & Vlachou, E. (2017). Augmented reality application to support remote maintenance as a service in the robotics industry. *Procedia CIRP, 63*, 46–51. https://doi.org/10.1016/j.procir.2017.03.154.

Oztemel, E., & Gursev, S. (2020). Literature review of Industry 4.0 and related technologies. *Journal of Intelligent Manufacturing*. https://doi.org/10.1007/s10845-018-1433-8.

Pan, S., Trentesaux, D., Ballot, E., & Huang, G. Q. (2019). Horizontal collaborative transport: Survey of solutions and practical implementation issues. *International Journal of Production Research*. https://doi.org/10.1080/00207543.2019.1574040.

Pan, S., Zhong, R. Y., & Qu, T. (2019). Smart product-service systems in interoperable logistics: Design and implementation prospects. *Advanced Engineering Informatics*. https://doi.org/10.1016/j.aei.2019.100996.

Poeppelbuss, J., & Durst, C. (2019). Smart service canvas - a tool for analyzing and designing smart product-service systems. In *Procedia CIRP*. https://doi.org/10.1016/j.procir.2019.04.077.

Schleich, B., Anwer, N., Mathieu, L., & Wartzack, S. (2017). Shaping the digital twin for design and production engineering. *CIRP Annals - Manufacturing Technology, 66*(1), 141–144. https://doi.org/10.1016/j.cirp.2017.04.040.

Shao, S., Xu, G., & Li, M. (2019). The design of an IoT-based route optimization system: A smart product-service system (SPSS) approach. *Advanced Engineering Informatics*. https://doi.org/10.1016/j.aei.2019.101006.

Söderberg, R., Wärmefjord, K., Carlson, J. S., & Lindkvist, L. (2017). Toward a Digital Twin for real-time geometry assurance in individualized production. *CIRP Annals - Manufacturing Technology, 66*(1), 137–140. https://doi.org/10.1016/j.cirp.2017.04.038.

Song, W. (2017). Requirement management for product-service systems: Status review and future trends. *Computers in Industry, 85*, 11–22.

Stark, J. (2015). *Product lifecycle management*. Cham: Springer. https://doi.org/10.1007/978-3-319-17440-2_1.

Takenaka, T., Yamamoto, Y., Fukuda, K., Kimura, A., & Ueda, K. (2016). Enhancing products and services using smart appliance networks. *CIRP Annals - Manufacturing Technology, 65*(1), 397−400. https://doi.org/10.1016/j.cirp.2016.04.062.

Tao, F., Cheng, J., Qi, Q., Zhang, M., Zhang, H., & Sui, F. (2017). Digital twin-driven product design, manufacturing and service with big data. *International Journal of Advanced Manufacturing Technology*, 3563−3576. https://doi.org/10.1007/s00170-017-0233-1.

Tao, F., & Qi, Q. (2019). New IT driven service-oriented smart manufacturing: Framework and characteristics. *IEEE Transactions on Systems, Man, and Cybernetics: Systems, 49*(1), 81−91. https://doi.org/10.1109/TSMC.2017.2723764.

Tao, F., Zhang, M., Liu, Y., & Nee, A. Y. C. (2018). Digital twin driven prognostics and health management for complex equipment. *CIRP Annals, 67*(1), 169−172. https://doi.org/10.1016/j.cirp.2018.04.055.

Uhlemann, T. H. J., Lehmann, C., & Steinhilper, R. (2017). The digital twin: Realizing the cyber-physical production system for industry 4.0. *Procedia CIRP, 61*, 335−340. https://doi.org/10.1016/j.procir.2016.11.152.

Vadde, S., Kamarthi, S. V., Gupta, S. M., & Zeid, I. (2008). Product life cycle monitoring via embedded sensors. *Environment Conscious Manufacturing*, 91−103.

Vazquez-Martinez, G. A., Gonzalez-Compean, J. L., Sosa-Sosa, V. J., Morales-Sandoval, M., & Perez, J. C. (2018). CloudChain: A novel distribution model for digital products based on supply chain principles. *International Journal of Information Management, 39*, 90−103. https://doi.org/10.1016/j.ijinfomgt.2017.12.006.

Verdugo Cedeño, J. M., Papinniemi, J., Hannola, L., & Donoghue, I. D. M. (2018). Developing smart services by internet of Things in manufacturing business. *DEStech Transactions on Engineering and Technology Research, 14*(icpr), 59−71. https://doi.org/10.12783/dtetr/icpr2017/17680.

Wang, Z., Zheng, P., Chen, C. H., & Khoo, L. P. (2018). A graph-based requirement elicitation approach in the context of smart product-service systems. In *48th international conference on computers and industrial engineering, Auckland, New Zealand*.

Watanabe, K., Okuma, T., & Takenaka, T. (2020). Evolutionary design framework for Smart PSS: Service engineering approach. *Advanced Engineering Informatics*. https://doi.org/10.1016/j.aei.2020.101119.

Weiß, P., Kölmel, B., & Bulander, R. (2016). Digital service innovation and smart technologies: Developing digital strategies based on industry 4.0 and product service systems for the renewal energy sector. In *Proceedings of the 26th annual RESER conference, Naples, Italy* (pp. 274−291).

Wiesner, S., Hauge, J., Haase, F., Thoben, K., Wiesner, S., Hauge, J., ... Haase, F. (2017). *Supporting the requirements elicitation process for cyber-physical product-service systems through a gamified approach to cite this version : HAL Id : hal-01615698 supporting the requirements elicitation process for cyber-physical product-service systems*, 0−8.

Wilding, R., & Juriado, R. (2004). Customer perceptions on logistics outsourcing in the European consumer goods industry. *International Journal of Physical Distribution and Logistics Management*. https://doi.org/10.1108/09600030410557767.

Zhang, Y., Liu, S., Liu, Y., & Li, R. (2016). Smart box-enabled product−service system for cloud logistics. *International Journal of Production Research, 54*(22), 6693−6706. https://doi.org/10.1080/00207543.2015.1134840.

Zhang, H., Qin, S., Li, R., Zou, Y., & Ding, G. (2020). Environment interaction model-driven smart products through-life design framework. *International Journal of Computer Integrated Manufacturing*. https://doi.org/10.1080/0951192X.2019.1686176.

Zheng, P., Chen, C.-H., & Shang, S. (2019). Towards an automatic engineering change management in smart product-service systems—A DSM-based learning approach. *Advanced Engineering Informatics, 39,* 203—213. https://doi.org/10.1016/j.aei.2019.01.002.

Zheng, P., Lin, T. J., Chen, C. H., & Xu, X. (2018a). A systematic design approach for service innovation of smart product-service systems. *Journal of Cleaner Production, 201*(August), 657—667. https://doi.org/10.1016/j.jclepro.2018.08.101.

Zheng, P., Lin, Y., Chen, C.-H., & Xu, X. (2018b). Smart, connected open architecture product: An IT-driven co-creation paradigm with lifecycle personalization concerns. *International Journal of Production Research, 0*(0), 1—14. https://doi.org/10.1080/00207543.2018.1530475.

Zheng, M., Ming, X., Wang, L., Yin, D., & Zhang, X. (2017). Status review and future perspectives on the framework of smart product service ecosystem. *Procedia CIRP, 64,* 181—186. https://doi.org/10.1016/j.procir.2017.03.037.

Zheng, P., Wang, Z., & Chen, C. (2019). Industrial smart product-service systems solution design via hybrid concerns. *Science, 00.* https://doi.org/10.1016/j.procir.2019.02.129.

CHAPTER 9

Toward sustainable smart product-service systems

Contents

9.1 Two directions for promoting sustainability	203
9.2 Fundamentals of sustainable smart PSS (SSPSS)	205
9.2.1 Definition	205
9.2.2 Key features	206
9.2.3 Response to environmental, economic, and social sustainability concerns	207
9.3 Systematic framework for developing SSPSS	208
9.4 A four-phase PDCA procedure in the cyber space	210
9.4.1 Core: context awareness	210
9.4.2 Plan: requirement elicitation (RE)	212
9.4.3 Do: solution recommendation (SR)	213
9.4.4 Check: solution evaluation (SE)	215
9.4.5 Adjust: knowledge evolvement (KE)	216
9.5 Case study	218
9.6 Chapter summary	225
References	226

Sustainable development is the utmost, recognized concern for today's industrial companies. It is elaborated into three aspects, i.e., environmental sustainability (reduce water/air pollution, fuel consumption, carbon emission), economic sustainability (allowing a modification/upgrade of components, cutting down logistics/transportation cost), and social sustainability (customer loyalty, shared value, improvement/enhancement to human well-beings). Toward these sustainability goals under the context of Smart PSS, a systematic framework with detailed procedures is proposed to prescribe the sustainable development process.

9.1 Two directions for promoting sustainability

Replying to the appeal of "doing more with less material" in a circular economy (Westkämper, Alting, & Arndt, 2000), industrial companies have proposed multiple types of reversible strategies to promote sustainability throughout the whole product lifecycle. These strategies include redesign,

---

Smart Product-Service Systems
ISBN 978-0-323-85247-0
https://doi.org/10.1016/B978-0-323-85247-0.00004-9

© 2021 Elsevier Inc.
All rights reserved.

203

remanufacturing, reuse, and recycle (i.e., 4R). As shown in Fig. 9.1, they formed a circular system manner. Aim to mitigate environmental impact and reduce unrenewable resource consumptions, redesign bridges the end-product and customer's experience in the usage stage with an inverse-design principle; remanufacturing restores a used product to equivalent or better performance of a new product; without changing the original state, reuse allows additional lifecycle cycles for several components or the whole product in an alternative application scenario; recycle extracts raw materials and usable components from the discarded product, which ultimately closes the circular loop.

Beyond "consuming less" with reversible strategies, another direction is "doing more" with the extended lifespan, which coincides with the concept of product-service system (PSS). With bundled customized services and reconfigurable products, customers' novel requirements can be met, and they are allowed to proceed to use the product beyond the originally designed lifespan. After the service innovation and product upgrade, the achieved functionalities are sometimes far beyond the originally designed propose. It also offers possibilities to the customers to use/reuse the product in another using scenario and hence achieves an additional lifespan.

Note that the circular system provides new revenue potentials and cost reductions to PSS (Michelini, Moraes, Cunha, Costa, & Ometto, 2017), meanwhile, Smart PSS showed high self-adaptability and built-in-flexibility that can implement smart circular effectively (Zheng, Lin, Chen, & Xu, 2018a). The overlapping of these two ideations, named Sustainable Smart PSS (SSPS), is emerging. Promoting sustainability in a holistic manner, it will fully exploit the user-generated data and product-sensed one, and perform sustainable reuse, reconfigure, maintenance, and recycling processes in the whole lifecycle stages in a smarter and more cost-efficient manner.

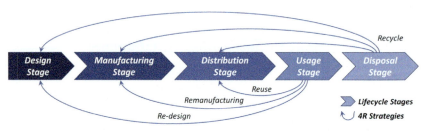

Figure 9.1 4R strategies in product lifecycle stages.

9.2 Fundamentals of sustainable smart PSS (SSPSS)

9.2.1 Definition

As demonstrated in Fig. 9.2, the basic concept of SSPSS can be seen as the trinary overlapping area of product-service systems, sustainable strategy, and smart technology. The formal definition of SSPSS can be given as: "a type of product-service systems empowered by the cutting-edge operational technology, information and communication technologies, and artificial intelligence, consisting of smart, connected products and their generated services as the solution bundle, to meet individual stakeholder's need in a sustainable manner." It can be further elaborated into three perspectives.

- PSS perspective

    SSPSS follows a value co-creation paradigm. However, via open-source and open-architecture, the openness of software and hardware in SSPSS is enhanced. Meanwhile, through a service-based incentive mechanism, the involvement of its massive users is promoted. Therefore, a user-centered open innovation is achieved, which continuously delivers value in the extended even circular lifecycle of SSPSS.

- Sustainable strategy perspective

    By better reallocating physical resources (e.g., components, material, energy) and cyber resources (e.g., empirical knowledge, historical records), SSPSS achieves additional product lifespan with a reversible and cost-efficient manner. With ever-evolving manners, it aims to enable and maintain a long-lasting customer relationship.

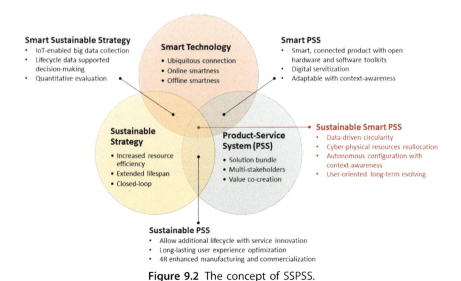

Figure 9.2 The concept of SSPSS.

- Smart technology perspective

   Via IoT infrastructure and big data analytics manner, lifecycle big data can be collected and transmitted, and massive semantic information can be retrieved in SSPSS. With multidisciplinary domain-specific knowledge and common sense stored in its knowledge base, it will self-learn the context information in the surrounding circumstance, and then self-configure itself under various perceived contexts, so as to achieve better performance.

9.2.2 Key features

According to the abovementioned definition, four key features in SSPSS can be summarized. In SSPSS, the essence is *data-driven circularity*, the methodology is *cyber-physical resource reallocation*, the manifestation is *autonomous configuration with context awareness*, and the motivation is *user-oriented long-lasting evolving*.

Data-driven circularity is in line with Data-Information-Knowledge-Wisdom (DIKW). Through IoT-enabled sensing devices (e.g., RFID, smart sensors) to collect product-sensed data, and social sensors (e.g., event-listener, webpage-crawling tools) to collect user-generated data, lifecycle data in whole stages is obtained as the data resources. With data analytics models (e.g., unsupervised clustering, supervised regression and classification) and domain-specific knowledge representation models (e.g., UML diagram, domain ontology, knowledge graph), the status information of SSPSS (e.g., reusability, reconfigurability) and some relevant enablers and ecosystems (e.g., availability of third-party services, the capacity of transportation) are mined, integrated and traced (Liu, Zhang, Zhang, & Zheng, 2019). Rules and empirical knowledge for lifecycle management can be further extracted, thereby supporting decision-making for sustainability (e.g., optimization of remanufacturing scheduling, upgrade of recycling serviceability) with a more solid foundation but shorter latency (Zhang, Ren, Liu, Sakao, & Huisingh, 2017).

Cyber-physical resource reallocation intends to promote sustainability in both physical and cyber spaces. Via reversible strategies (i.e., 4R), tangible resources (e.g., raw materials, reusable components) in SSPSS are almost reallocated in the smart circular system (Alcayaga, Wiener, & Hansen, 2019). Meanwhile, intangible resources (e.g., obtained sensing dataset, annotated context information, and concluded empirical constraints) should be also reallocated. In upgrading products and/or generating services, a knowledge management system should reuse or reorganize previous concepts and propositions, so as to rapidly provide innovative but cost-effective solutions (i.e., transfer of knowledge Li, Jiang, Guan, Li, & Wang, 2019).

Autonomous configuration with context awareness indicates the maximum rank of connectedness and smartness (Lee, Bagheri, & Kao, 2015). Based on the relevant domain knowledge and common sense, context information in the lifecycle stages is automatically perceived. Responding to perceived physical/social/user/operational context features, an informed circularity decision can be self-made to rapidly self-configure product and/or service components, even in real-time.

User-oriented long-lasting evolving is essential to meet users' rising requirements, so as to establish and cultivate a long-lasting relationship (Liu, Song, & Han, 2020). Achieved by open-architecture hardware and open-source software, the degree of innovation flexibility becomes much higher. In this situation, in the extended lifespan of SSPSS, end users can take the role to direct the development process. Therefore, the finally achieved functionalities and delivered values will largely deviate from initially designed ones (Zheng, Lin, Chen, & Xu, 2018b). Meanwhile, the reversed process, which starts from the latter usage and/or disposal stages and ends at the former design, manufacturing, and/or distribution stages, could be the majority in SSPSS development.

9.2.3 Response to environmental, economic, and social sustainability concerns

Compliant with the United Nations' 2030 Agenda for Sustainable Development blueprint, especially the Goal 9 to "build a Build resilient infrastructure, promote inclusive and sustainable industrialization and foster innovation," SSPSS is capable to respond to the expectations on environmental, economic, and social sustainability concerns.

- Environmental aspect

    SSPSS follows the environmental management system strategies of "reduce, reuse and recover," and the eventual goal is to scale down environmental impacts and realize sustainability appealed in the circular economy. The environmental concerns can be readily addressed by extended lifecycle, better resource use efficiency, and the circulate lifecycle. However, appropriate waste electrical and electronic equipment (WEEE) treatment should be handled resulting from the ever-increasing smart products utilized.

- Economic aspect

  In SSPSS, digitalization capabilities enable the value creation/capture, proposition, and delivery in a context-aware and cost-effective manner. Value is created/captured in a data-driven manner from those massive stakeholders and their end smart, connected products in a co-creation process. Meanwhile, it reflects the sustainable, digital and servitized value proposition, by not only addressing the three big transformations among today's industrial companies internally, but also to meet personalized needs with sustainable, smart, on-demand solutions. Furthermore, value is delivered in a Software-as-a-Service (SaaS) manner, where stakeholders have ubiquitous access to the cyber-physical resources shared, reused, remanufactured, and recycled in a sustainable way.

- Social aspect

  SSPSS is highly consistent with the idea of Open Innovation 2.0, where various stakeholders contribute in the value co-creation process to meet their demands. Meanwhile, it aims to make sustainable prosperity and enhancements in human well-being by leveraging some disruptive smart enabling technologies to provide product-service offerings.

9.3 Systematic framework for developing SSPSS

Based on the abovementioned features, Fig. 9.3 depicts the ultimate goal of the development of SSPSS.

In accordance with the engineering lifecycle, the horizontal arrow indicates the implementation stages of PSS. Referring to the reversible strategies in lifecycle management, physical resources should be reallocated holistically in the involved stages, which includes smart redesign (e.g., upgrade of product-service modules), smart remanufacturing (e.g., prognostics and health management), smart reuse (e.g., smart rental/sharing system), and smart recycling (e.g., smart WEEE management). The conduction of each strategy is no longer isolated in one or a few lifecycle stages but should be informed and guided by the data and knowledge accumulated in the circular life span so as to achieve higher context awareness and cost-efficiency.

The vertical arrow represents the DIKW flow in the digital servitization process, which is supported by the cutting-edge ICT infrastructure (e.g., cloud-edge computing), digital technologies (e.g., digital twin), and AI

Toward sustainable smart product-service systems 209

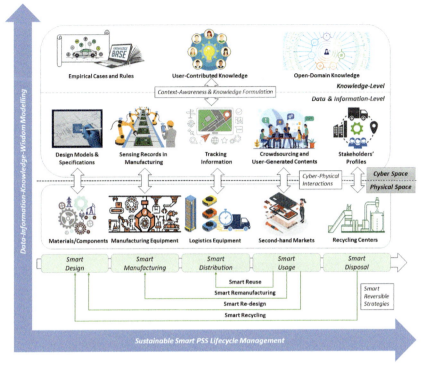

Figure 9.3 Systematic framework for SSPSS development.

techniques (e.g., deep learning, knowledge graph). First, velocity, veracity, value, volume, variety (5V) data is collected via various sources in a heterogeneous manner, including both user-generated data and product-sensed one. Second, unified information modeling should be established to manage the cleaned data with big data analytic methods. Third, valuable knowledge should be extracted from the information models, and the domain ontologies and specific knowledge graphs should be constructed for knowledge representation and reasoning purposes in a cognitive (semantic) manner. At last, knowledge-driven solutions for implementing reversible strategies should be proposed/recommended as business intelligence.

Meanwhile, the cyber-physical interactions (i.e., digital twin) in SSPSS development can be cost-effectively established between the physical and cyberspace, based on intelligent decision-making and real-time bidirectional communication between them (e.g., monitoring, control, and optimization).

9.4 A four-phase PDCA procedure in the cyber space

Existing studies have emphasized much on the sustainable concerns of the physical products/components, while little reported on the cyber space. Aiming to bridge the gap, a Plan-Do-Check-Adjust (PDCA) procedure as regulated in the ISO14001:2015 is provided as a phase-by-phase approach for the holistic development process of SSPSS, especially in the cyberspace. As shown in Fig. 9.4, the PDCA procedure is an iterative process that guides the stakeholders to achieve context-aware sustainable evolvement.

The flow of DIKW in the PDCA procedure is further prescribed in Fig. 9.5. Specifically, based on historical and real-time product-sensed data and user-generated one, the set of requirements and the set of solutions are mapped and evaluated. Four knowledge evolvement strategies are also proposed to update the rules and constraints in SSPSS development.

9.4.1 Core: context awareness

As the core and prerequisite of the PDCA procedure, context awareness intends to represent context information embedded in the large volume of product-sensed data and user-generated one. Regarding available sorts and contents of information that are capable to be cost-effectively obtained, four classes of context features can be categorized in SSPSS development (Li et al., 2020): (1) *Physical context* (environmental information), (2) *Social context* (status of nearby components and systems), (3) *User context* (user

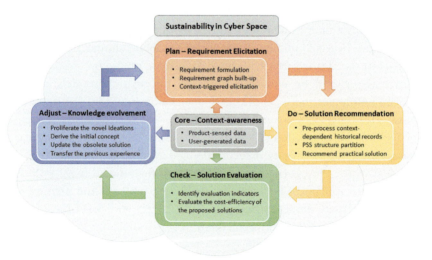

Figure 9.4 Four-phase PDCA procedure in the cyber space.

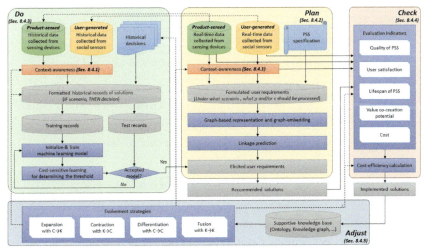

Figure 9.5 DIKW in the PDCA procedure.

information and interactions between users and PSS), and (4) *Operational context* (operational information of product/service components). Examples of context features are listed in Table 9.1.

Based on the aforementioned context features, scenarios embedded in collected datasets are encoded with a key-value modeling approach. For each context feature c_i in the k-element-set $C = \{c_i\}_k$, a value v_i is correspondingly filled. As illustrated in Fig. 9.6, a k-dimensional-vector, $sn = [v_1, v_2, ..., v_k] \in ¡^k$, is hence formed for representing the scenario. Since the datasets in the process of SSPSS development are heterogeneous, Table 9.2 also presents some frequently used data analytics techniques in determining the context value.

Table 9.1 Four classes of context features.

Classes	Exemplary context features
Physical	Date; time; location; direction; temperature; humidity; odor; air/water quality; weather … …
Social	Peer products; third-party service provider; available recycler; second-hand market orders … …
User	Demographics; mood/health; experience; preference/habit; usage type … …
Operational	Version; maintenance history; Portability/Wearability; computing power … …

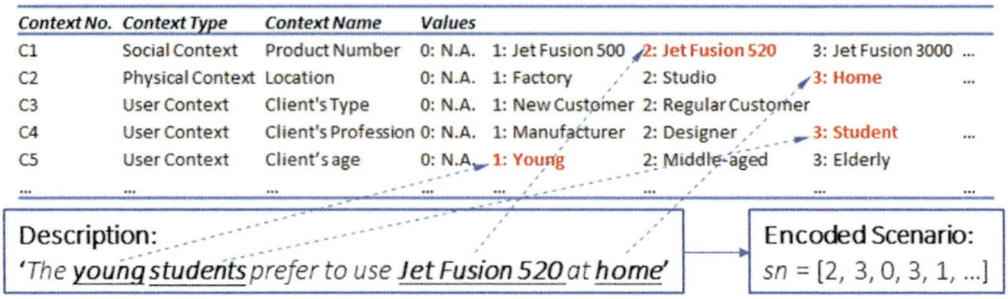

212 Smart Product-Service Systems

Context No.	Context Type	Context Name	Values			
C1	Social Context	Product Number	0: N.A.	1: Jet Fusion 500	2: Jet Fusion 520	3: Jet Fusion 3000 ...
C2	Physical Context	Location	0: N.A.	1: Factory	2: Studio	3: Home ...
C3	User Context	Client's Type	0: N.A.	1: New Customer	2: Regular Customer	
C4	User Context	Client's Profession	0: N.A.	1: Manufacturer	2: Designer	3: Student ...
C5	User Context	Client's age	0: N.A.	1: Young	2: Middle-aged	3: Elderly
...	...	...	...	...		...

Description:
'The **young students** prefer to use **Jet Fusion 520** at **home**'

Encoded Scenario:
$sn = [2, 3, 0, 3, 1, ...]$

Figure 9.6 Encoding scenarios using context features.

9.4.2 Plan: requirement elicitation (RE)

As the first phase in the PDCA procedure, RE intends to distinguish and represent end users' potential requirements. A data-driven manner can be leveraged to extract implicit user requirements. The elicited requirements will guide the subsequent product/service innovation.

Two data resources can be leveraged for RE, namely, user-contributed feedbacks in the website (e.g., comments, scores/ratings, Q&A in forums), and signal data collected from sensing devices (e.g., position, humidity, acceleration, velocity, temperature). In order to concern the context-dependency in the dataset, a format can be used for RE, i.e., "given a certain scenario, what product structures and/or service modules should be changed, updated, reused, and/or recycled" (Wang, Chen, Zheng, Li, & Khoo, 2019a). Accordingly, a requirement can be expressed with a three-element-tuple, i.e., req $= \langle \{p\}, \{s\}, sn \rangle$, where $p \in P$ and $s \in S$ separately

Table 9.2 Data analytics techniques for determining context values.

Data sources & types	User-generated		Product-sensed	
	Structured text	Natural language	Numerical value	Numerical value
Techniques	Table headers and elements Formal concept analysis Schema-based annotation Predefined template	Keyword extraction Named-entity recognition Syntax analysis Sentiment analysis	Use domain knowledge Use common sense Fuzzy rules Rough sets	Pattern recognition Use domain knowledge Fuzzy rules Rough sets ... ...

indicate decomposed product and service components in the SSPSS (i.e., $PSS = P \cup S$, $P \cap S = \varnothing$), and $sn \in SN$ is presented with a k-dimensional vector. Therefore, RE can be regarded as an exploration of the cooccurrence relationship among products, services and scenarios. In this situation, the graph-based approach is appropriate for achieving the goal. A requirement graph, $RG = \langle V, E \rangle$, is established, where its set of vertexes is the union of products, services, and scenarios, i.e., $V = P \cup S \cup SN$. The set of edges E represents the cooccurrence relationship between nodes (i.e., in a piece of comment, two product/service/context entities appeared simultaneously). In this way, RG can be incrementally extended, with new nodes of products, services and scenarios generated in the development process.

Based on RG, eliciting novel requirements in SSPSS development can be seen as the graph-based linkage prediction. With some graph-embedding algorithms (e.g., DeepWalk, SkipGram), when perceiving a specific scenario, several $p\text{-}sn$ and/or $s\text{-}sn$ edges in RG which achieve the highest appearance probabilities can be selected. They will form an explicitly expressed requirement, and guide the subsequent user-oriented upgrade (Wang, Chen, Zheng, Li, & Khoo, 2019b).

9.4.3 Do: solution recommendation (SR)

Since RE reflects directions for upgrading products from users' perspectives, instead of other stakeholder's perspectives (e.g., designers, manufacturers, service providers, operators, and recyclers), it will omit some restrictions in the practical. Thereby, SR, as the second phase in the PDCA procedure, is carried out to supply a more feasible and practical solution by analyzing historical records generated during SSPSS development.

Historical records are some empirical but valuable knowledge, like successful problem-solving documents/lesson-learned cases. Their contents can be simplified to "IF a scenario occurs, THEN change, update, reuse, and/or recycle the selected product/service components." Corresponding to this IF-THEN format, a historical record contains two elements, namely, $rec = \langle sn, d \rangle$, where sn annotates a piece of historical scenario using a k-dimensional vector encoded with context features listed in in Table 9.1 and Fig. 9.6, and $d = \langle \{p\}, \{s\} \rangle$ represents corresponding selections of product and/or service components.

Apparently, when some previous problem-solving contexts reappear, several accumulated knowledge will be straightforwardly reused, and feasible solutions of changing, updating, reusing, and/or recycling some formerly chosen components in the specific case will be rapidly offered. However, when a novel scenario occurs is perceived and an undiscovered value combination of context features is discovered, former solutions should be modified before recommending to stakeholders. To achieve this with higher automation and shorter latency, machine-learning manners can be leveraged, for instance, Naïve Bayes, Random Forest, and Support Vector Machine (SVM). Specifically, to train a prediction model with massive historical records, a matrix is firstly established. This matrix can be partitioned into two subsets, namely, scenario set (context feature values) and decision set (selected product and/or service components). In model training, the probability of selecting each product and service component in the solution is respectively forecasted. Considering the error in classification, the performance of the machine-learning manner is evaluated and optimized. After that, aiming to decide the threshold in the decision-making, namely, the minimum possibility for component selection in the output solution, the teaching cost for determining the boundary region in the classification is estimated via a cost-sensitive training approach (Zheng, Chen, & Shang, 2019).

Meanwhile, considering an ever-evolving SSPSS with a more complicated structure, the number of product/service components will be continuously increased, which simultaneously brings exponentially growing combinations of decisions. In this situation, if only a relatively small dataset can be collected for model training, the prediction precision may deteriorate. To mitigate this, in the learning process, clustering approaches are leveraged to reduce the dimension of the prediction model. Specifically, rely on the total records, a cooccurrence matrix is established in the first step. Each slot in this matrix represents the relative cooccurrence frequency between two components. With modularity calculation (e.g., community-partitioning algorithms; Blondel, Guillaume, Lambiotte, & Lefebvre, 2008), communities are discovered and hence partitioned. With partition results, the decision set is also renewed to a modular level. Therefore, in operating the machine-learning approach mentioned above, the dimension is largely reduced to the number of clusters. It will raise the feasibility of the proposed data-driven SR in the PDCA procedure.

9.4.4 Check: solution evaluation (SE)

To keep competitive in the market, in the development of SSPSS, it is unwise to chase better performance, longer lifespan, and/or higher user satisfaction aimlessly and endlessly. Only more cost-effective solutions in product/service innovation can be adopted. Thereby, as the third phase in the PDCA procedure, SE will measure and optimize the cost-efficiency of proposed solutions, aiming at achieving a better balance between the cost and benefit.

Five criteria are proposed for SE (Shen, Erkoyuncu, Roy, & Wu, 2017), i.e., (1) maximizing the quality of PSS (Q); (2) maximizing the user satisfaction (US); (3) maximizing the lifespan of PSS (LS); (3) maximizing value co-creation potential (VC); and (5) minimizing the cost for evolvement (C). These criteria consider value-proposition capability, PSS-customer relations, and the total cost in developing SSPSS, which are quantitatively measured with Eqs. (9.1)−(9.5).

$$Q = 1 - \alpha_1 \sum_{PSB} k(performance - goal)^2 \tag{9.1}$$

$$US = \frac{\alpha_2}{|PSB|} \sum_{PSB} \left(\overline{rate} - \overline{rate_0} \right) \tag{9.2}$$

$$LS = \alpha_3 \frac{\overline{lifespan_{PSB}} - \overline{lifespan_0}}{\overline{lifespan_0}} \tag{9.3}$$

$$VC = \frac{\alpha_4}{|PSB|} \sum_{PSB} Score_{potential} \tag{9.4}$$

$$C = \alpha_5 \sum_{PSB} (C_P + C_S + C_H + C_I) \tag{9.5}$$

Q in Eq. (9.1) is computed as the rest quality after deducting the quality loss. Defined by Taguchi (1995), Q is evaluated with the normalized deviations for desired goals, and computed on each product-service bundle (PSB).

US in Eq. (9.2) annotates the promotion of US in the recommended solution. On each PSB, US is quantified by the user-generated online feedbacks and/or ratings, via time-series analysis and sentiment analysis.

LS in Eq. (9.3) evaluates the extendibility of lifespan. The extended lifespan can be estimated using historical lifecycle data, when a specific solution is put into action.

VC in Eq. (9.4) indicates some abilities possessed in product–service bundles that can enhance the innovation (like openness, smartness, and connectedness). Scored with some predefined models (e.g., 5C level architecture Lee et al., 2015), the higher metric value indicates higher availability in value co-creation.

As for C in Eq. (9.5), it is the total cost consumed for implementing the solution, which is constituted of four aspects, namely, cost for physical resources C_P, service-relevant processing C_S, human resources C_H, and intellectual resources C_I. These costs can be obtained from multiple aspects of stakeholder.

Besides, $\alpha_1-\alpha_5$ in Eqs. (9.1)–(9.5) are five fixed normalization coefficients. They will coordinate Q, US, LS, VC, and C to an approximate order of magnitude.

After the evaluation of the abovementioned five criteria, the overall cost-efficiency (CE) of a solution generated in the previous PDCA procedure is hence computed:

$$CE = \frac{w_1 * Q + w_2 * US + w_3 * LS + w_4 * VC}{C} \qquad (9.6)$$

where w_1-w_4 are four dynamic weights that are self-adjusted by the user preference. Apparently, for a bunch of candidate solutions, one with the highest CE will be adopted and implemented in the development of SSPSS.

9.4.5 Adjust: knowledge evolvement (KE)

After an innovative product–service solution has been carried out, product and/or service components will be partly or totally changed, updated, reused, and/or recycled. Accordingly, the relevant knowledge acquired in all the lifecycle stages also needs evolvement, for example, modifying the design principle, revising the manufacturing craft, changing the logistic restraint, adjusting the usage manner, and/or updating the dismantling information. Therefore, as the last phase in the PDCA procedure, KE intends to handle the revisions and ensures the coherence in the knowledge base in the long-lasting SSPSS development, thus closing the PDCA procedure.

Inspired by four patterns (knowledge expansion, knowledge contraction, knowledge differentiation, and knowledge fusion) discovered in the long-term evolvement (Li et al., 2017, 2018) and correspondingly four operators (**C→K**, **K→C**, **C→C**, and **K→K**) claimed in Concept-

Knowledge (C-K) theory (Hatchuel & Weil, 2009), four heuristic strategies are designed to provoke KE in SSPSS development. Through these strategies, a knowledge management review is hence conducted to periodically and incrementally modify concepts and their relations stored inside the supportive knowledge base (e.g., domain ontology, knowledge graph).

- Knowledge expansion strategy: Proliferate novel ideations via $C \rightarrow K$ operator

  $C \rightarrow K$ operator represents an operation of rearranging concepts to construct a creative proposition (i.e., innovative knowledge). Following the ideation of this operator, a knowledge expansion strategy can be proposed. Rely on the implemented innovative solution, a "knowledge family" is established in SSPSS. Specifically, via default inference, concepts leveraged in these solutions can be linked up to form a group of proliferated propositions, when there is no observed logical conflict to the current knowledge items.

- Knowledge contraction strategy: Update obsolete solutions via $K \rightarrow C$ operator

  As a reversed operation of $C \rightarrow K$ operator, $K \rightarrow C$ operator intends to suggest some novel concepts. It imports relevant attributes to the concept space from the extant knowledge. During this process, the logical coherence should be guaranteed. Therefore, in $K \rightarrow C$ operator, out-of-date solutions that use former concepts will be correspondingly renewed, and the opportunity for utilizing the solution in the follow-up development process of SSPSS is decreased.

- Knowledge differentiation strategy: Derive initial concepts via $C \rightarrow C$ operator

  Similar to $K \rightarrow C$ operator, $C \rightarrow C$ operator also introduces innovative attributes to suggest some new concepts. However, it intends to derive the content of a generic concept and specify its scope in a new scenario. Based on this idea, with the consideration of some novel context features or abnormal scenarios, the knowledge differentiation strategy will search a derived concept in highly relevant domains. Therefore, some alternative options for self-adaptation can be automatically provided for SSPSS development.

- Knowledge fusion Strategy: Transfer previous experience with $K \rightarrow K$ operator

  Through all typical types of logic reasoning (classification, abduction reasoning, deduction reasoning, inference), $K \rightarrow K$ operator aims to establish a logical relationship between a newly achieved piece of knowledge and an extant one in SSPSS. Relying on the logical chain/network constructed in the fusion process, it is hence available to reuse previously generated empirical knowledge in some other problem-solving contexts. A partly or totally transferred solution under a new scenario will be accordingly generated.

9.5 Case study

To demonstrate the application of the four-phase procedure, a smart 3D printer is presented as a case study. Regarded as an ecofriendly product, good sustainability of 3D printer achieved in the rapid reconfiguration and remanufacturing with recyclable raw materials and reusable product components. Meanwhile, enabled by the digital twin technique, several tailored added-on services, for instance, remote controlling/monitoring, printing-maintenance joint scheduling, and consumable management, can be bundled to the 3D printer, which shows great potentials to realize an SSPSS. However, due to the inadequate utilization of the context-rich lifecycle information and knowledge, the current 3D printer cannot be regarded as sustainable in the cyberspace. To achieve the goal for a cyber-physical smart 3D printer, the proposed framework and the four-phase PDCA procedure was conducted, as shown in Fig. 9.7.

Since performing every aspect along all the lifecycle stages is too complicated, in this case, the PDCA procedure implementing the smart

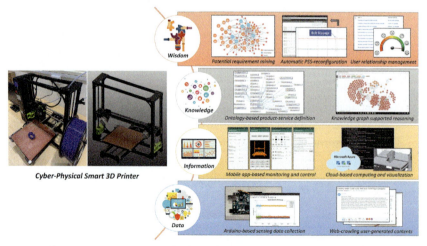

Figure 9.7 A prototype of cyber-physical smart 3D printer Platform.

redesign strategy was showcased. 20 essential product components and six service components were considered in this case, as indexed in Table 9.3. To realize context awareness, as listed in Table 9.4, seven context features were chosen, regarding available sensing datasets, and suggestions provided by 3 3D printing experts. These experts were also invited to check the generated solutions and verify the proposed framework.

To implement RE, 85 newly generated threads (2019.06−2019.08) were collected from *forum.lulzbot.com*, which is the official online platform for the users of LulzBot 3D printers. With one-hot encoding, keywords in each thread were extracted. Mapping to context features predefined in Table 9.4, an encoded scenario is hence formed for each thread. Meanwhile, words indicating the product and service components were discovered, thus formulating the three-element-tuple $req = \langle \{p\}, \{s\}, sn \rangle$. Based on these tuples, a requirement graph was constructed with the established edges of *p-s*, *p-p*, *p-sn*, *s-s*, and *s-sn*. As illustrated in Fig. 9.8, it depicted the interrelations among all possible usage scenarios (as shown in red nodes) and product/service components (as shown in orange and blue nodes).

As reported in Table 9.5, predicted by SkipGram algorithm, scenarios with top three highest occurrence frequency, as well as product and service components with top five highest appearance probabilities, are fetched in context-aware RE, so as to represent user requirements (Wang et al, 2019a, 2019b). For instance, under a perceived scenario of $[-1, -1, 0, 1, 0, 0, 2]$, which means "Low temperature for certain filaments" (i.e., $c1$: *Nozzle Temperature is below*

Table 9.3 A list of product and service components in the 3D printer platform.

Product components		
$p1$: Nozzle	$p8$: Extruder gear	$p15$: Thermistor
$p2$: LCD screen	$p9$: Z-axis lead screw	$p16$: Heat break
$p3$: X tension belt	$p10$: X stepper motor	$p17$: Heat sink
$p4$: Y tension belt	$p11$: Y stepper motor	$p18$: Nozzle fan
$p5$: PEI surface print bed	$p12$: Z stepper motor	$p19$: Part fan
$p6$: Rambo board	$p13$: Extruder stepper motor	$p20$: Filament
$p7$: Bearing	$p14$: Heat bed cable	
Service components		
$s1$: Parameter configuring	$s3$: Quality checking	$s5$: Inventory management
$s2$: Printing tracking	$s4$: Maintenance scheduling	$s6$: Payment selection

Table 9.4 A list of context features.

Context feature	Context class	Context values			
$c1$: Nozzle temperature	Physical context	−1: <170°C	0: 170−220°C	1: >220°C	
$c2$: Extrusion speed	Physical context	−1: <40 mm/s	0: 40−60 mm/s	1: >60 mm/s	
$c3$: Layer height	Physical context	−1: <0.14 mm	0: 0.14−0.38 mm	1: >0.38 mm	
$c4$: Clogging	Operational context	/	0: No issue	1: Nozzle clogged	
$c5$: String	Operational context	/	0: No issue	1: Filament stringing	
$c6$: Second-hand status	Social context	/	0: Brand new	1: Second-handed	
$c7$: User type	User context	0: N.A.	1: Novice (<30 h)	2: Ordinary (30−100 h)	3: Expert (>100 h)

Toward sustainable smart product-service systems 221

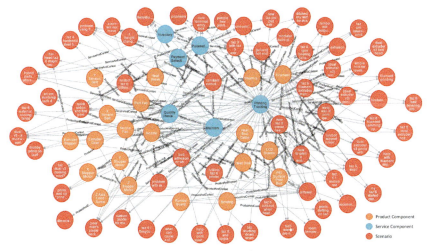

Figure 9.8 Requirement graph for 3D printer PSS.

Table 9.5 Top 3 RE results.

Requirements	Encoded *sn*	Description of *sn*	Predicted *p* and *s*	Probability
R1	[−1, −1, 0, 1, 0, 0, 2]	Low temperature for certain filaments	p20: Filament	0.95
			p18: Nozzle fan	0.93
			s1: Parameter Configuring	0.85
			p15: Thermistor	0.81
			s4: Maintenance scheduling	0.78
R2	[0, 0, 1, 0, 1, 0, 1]	Shifting layers with poor support	s1: Parameter Configuring	0.97
			p5: PEI surface print bed	0.87
			p20: Filament	0.80
			p4: Y tension belt	0.72
			p3: X tension belt	0.72
R3	[0, −1, 0, 0, 0, 1, 2]	Extrusion failure after repair	s4: Maintenance scheduling	0.94
			p20: Filament	0.92
			p1: Nozzle	0.90
			s3: Quality checking	0.77
			p8: Extruder gear	0.72

170°C, *c2*: *Extrusion Speed is less than* 40 mm/s, *c3*: *Layer Height is between* 0.14 and 0.38 mm, *c4*: *Nozzle Clogged*, *c5*: *No filament stringing issue*, *c6*: *Brand new printer* and *c7*: *Ordinary user*), requirement *R1* was discovered. According to corresponding discussion threads in the platform, three product components, i.e., *p20*: *Filament*, *p18*: *Nozzle Fan*, and *p15*: *Thermistor*, and two service components, i.e., *s1*: *Parameter Configuring* and *s4*: *Maintenance Scheduling*, were the most mentioned. Therefore, changing/upgrading these components under the selected scenario was extracted as the user requirement.

To handle the aforementioned requirements, 1802 records (including repairing, replacing, upgrading logs) of the selected model of 3D printers were collected for SR. These records were preprocessed to set up the training model. As illustratively presented in Table 9.6, a scenario is encoded with seven features proposed in Table 9.4, and the decision set listed the selected combination of repaired, replaced, and/or upgraded product and/or service components under the encoded scenario.

To reduce the dimension in the predicting, 20 product components and six service components were partitioned into five clusters, through cooccurrence frequency analysis and community-partitioning processing, as shown in Table 9.7. Then, on the training dataset, a random forest model was trained using 10-fold cross-validation. After that, the model was utilized to generate possible repairing, replacing, and/or upgrading solutions for solving the discovered requirements presented in Table 9.5 (Zheng et al., 2019), as shown in Table 9.8. For instance, in the recommended solution *So1* to solve *R1* (*Low temperature for certain filaments*), two product components, *p15*: *Thermistor* and *p20*: *Filament* were predicted to be replaced. Another two product components, *p16*: *Heat break* and *p17*: *Heat*

Table 9.6 Preprocessed historical records.

Record no.	Encoded scenario set							Decision set (repaired, replaced, and/or upgraded product and service components)
	c1	*c2*	*c3*	*c4*	*c5*	*c6*	*c7*	
1	0	0	0	0	1	0	2	*p1, p8, p14, p15, s1, s4*
2	0	0	−1	0	0	1	1	*p7, p9, p12, p19, s2, s4*
3	−1	0	−1	1	1	0	1	*p5, p7, p8, p9, p12, p13, s2, s3, s4*
4	1	0	0	1	0	1	2	*p5, p14, p18, p19*
5	0	1	0	1	0	0	2	*p5, p14, s1, s4*
...	...	...	...	...	...	...	...	...

Table 9.7 Cluster division in the 3D printer.

Cluster no.	Contained product and service components	Descriptions
cl1	*p1, p5, p7, p8, p13, p14, p18, p19, s4*	Extruding module
cl2	*p2, p6, s2*	Printing tracking module
cl3	*p3, p4, p9, p10, p11, p12, s3*	Movement module
cl4	*p15, p16, p17, s1*	Heating module
cl5	*p20, s5, s6*	Consumable management module

sink were predicted to be changed. Three service components, *s1: Parameter Configuring, s5: Inventory Management,* and *s6: Payment Selection* were also predicted to be upgraded.

Aiming to measure the CE of generated solutions, SE in the PDCA procedure was conducted. Five evaluation indicators in Eqs. (9.1)−(9.5) were evaluated with the experimental data collected from each evolved 3D printer prototype system. To maintain the information confidentiality, only finally scored results were presented. Raw data in the evaluation, including product/service components' specifications, cost/price/benefit, estimated lifespan, and the rating scores, was concealed. The dynamic weights w_1-w_4 in Eq. (9.6) were determined via an online 5-point Likert Scale-based questionnaire, which was conducted on a group containing 7 novice users ($c7 = 1$) and 11 ordinary users ($c7 = 2$). The weights were [0.57, 0.71, 0.89, 0.82] and [0.89, 0.84, 0.73, 0.59] separately.

Table 9.8 SR results for the elicited requirements.

Req.	Encoded *sn* [c1, c2, c3, c4, c5, c6, c7]	Probability of selection [P(cl1), P(cl2), P(cl3), P(cl4), P(cl5)]	Decision [cl1, cl2, cl3, cl4, cl5]	Repaired, replaced, and/ or upgraded *p* and *s*
R1	[−1, −1, 0, 1, 0, 0, 2]	[0.04, 0.11, 0.01, 0.74, 0.63]	[0, 0, 0, 1, 1]	*So1*: *p15, p16, p17, p20, s1, s5, s6*
R2	[0, 0, 1, 0, 1, 0, 1]	[0.17, 0.13, 0.72, 0.78, 0.24]	[0, 0, 1, 1, 0]	*So2*: *p3, p4, p9, p10, p11, p12, p15, p16, p17, s1, s3*
R3	[0, −1, 0, 0, 0, 1, 2]	[0.94, 0.00, 0.28, 0.20, 0.32]	[1, 0, 0, 0, 0]	*So3*: *p1, p5, p7, p8, p13, p14, p18, p19, s4*

224   Smart Product-Service Systems

As shown in Table 9.9, for ordinary users, the CE for generated solutions *So1* and *So3* were acceptable. Specifically, several product/service components in the heating module and consumable management modules, like *p15: Thermistor* and *p20: Filament*, can be changed to handle *low temperature for certain filaments (R1)*. Meanwhile, product/service components in the extruding module should be replaced/upgraded to handle *the extrusion failure after repair (R3)*. The effectiveness of two solutions was also validated by 3D printing experts invited in this case study. However, *So2* performed poorly in the evaluation of CE, caused by relatively high C in pursuing excessively high performance in Q and US. Therefore, the reconfiguration in *So2* should be reconsidered, before implementing it to the novice users. For instance, rethink the needfulness of each component that should be repaired, replaced, and/or upgraded, and optimize the final selected ones in the solution.

After the implementation of high CE solutions, KE in the PDCA procedure was to manage 3D printing knowledge with the proposed four evolvement strategies. For instance, in *So1, filament (p20)* was replaced. The relevant knowledge, *feed filament (p20) to the nozzle (p1)*, should be correspondingly modified. In this situation, $\mathbf{C} \rightarrow \mathbf{C}$ operator was operated on the concept of *filament*. Possessing the attribute of *melting temperature of 58° C*, a subconcept, *PCL filament* (annotated as *p20_1*), was discovered to extend *filament*. Based on the derived concept, $\mathbf{C} \rightarrow \mathbf{K}$ operator proposed an evolved knowledge, i.e., *feed PCL filament (p20_1) to the nozzle (p1) when the nozzle temperature is below 170° C (c1 = −1) and the user type is ordinary user (c7 = 2)*. Since no confliction to other pieces of 3D printing domain knowledge was observed, KE updated the former one in the follow-up knowledge-based procedure (i.e., conducted a $\mathbf{K} \rightarrow \mathbf{C}$ operator). Meanwhile, it constructed interrelations to other pieces of knowledge through $\mathbf{K} \rightarrow \mathbf{K}$ operator. Some more complicated propositions could be proposed, for instance, a piece of compound knowledge, *updating parameter configuring service (s1) for the ordinary user (c7 = 2) to change the nozzle temperature to below 170° C (c1 = − 1), when feeding PCL filament (p20_1) to the nozzle (p1)*.

Table 9.9 SE results on the generated solutions.

Solution no.	Evaluation indicators					Indicators' weights $[w_1, w_2, w_3, w_4]$	CE
	Q	US	LS	VC	C		
So1	0.92	0.76	0.75	0.63	2.8	[0.89, 0.84, 0.73, 0.59]	0.85
So2	0.98	0.96	0.36	0.55	3.8	[0.57, 0.71, 0.89, 0.82]	0.53
So3	0.82	0.96	0.53	0.51	2.4	[0.89, 0.84, 0.73, 0.59]	0.95

Accordingly, in the 3D printing ontology that supports the Smart 3D printer platform, KE created a sub-node of *PCL filament* and attached it to the extant node of *filament*. A new record, $rec = \{sn = [-1, 0, 0, 0, 0, 0, 2], d = \langle p1, p20\_1, s1 \rangle\}$, was also included in the analyzed historical datasets. When a new PDCA procedure started, flows in the first three phases (i.e., RE, SR, and SE) are impacted, thus guaranteeing the consistency in the knowledge base.

9.6 Chapter summary

Sustainable development has been a hot topic over the past few years, and ever-increasingly concerned by not only the government policymakers, but also modern companies to take proactive actions toward sustainability. Meanwhile, the rapid development of cutting-edge ICT and AI, as the smart enabling technologies, has ensured the digital servitization transition, where Smart PSS can be realized to meet individual customer needs readily. By nature, the "smart" balance between sustainable development and high value-added solutions may not necessarily conflict with each other, but to make decision-making better.

Motivated by that, this chapter aimed at discovering the core relationship between sustainability and Smart PSS, and further to prescribe the SSPSS development process. A systematic framework is introduced to utilize high value and context-rich information and knowledge during the development process. Inspired by the PDCA procedure regulated in ISO14001:2015, a four-phase process is further prescribed for the development process in the cyberspace. With the iterative flow of RE, SR, SE, and KE, it better supports decision-making and optimization in conducting reversible strategies on the cyber-physical resources of SSPSS. In a case study of redesigning a smart 3D printer platform, the proposed framework and the four-phase PDCA procedure achieved higher context awareness and cost-efficiency, proving their feasibility and advantages.

As an exploratory study, this chapter emphasized an overall development framework and a generic PDCA procedure. However, some detailed techniques, flowcharts, and algorithms are oversimplified. Regard this, future studies can concentrate on the following three perspectives:

(1) Include some incentive mechanisms collect more user-generated data, and adopt some machine-learning approaches that are not depended on massive training data, in order to mitigate the "cold start" issue for a new SSPSS;

(2) Upgrade the chosen data analytics manners and context awareness manners with some NLP techniques and image processing techniques, so as to include more sorts of available datasets in the proposed PDCA procedure;

(3) Import multidisciplinary open-domain knowledge and common sense to the knowledge base to better facilitate sustainable strategies under multiple scenarios in the development of SSPSS, which will enhance the logical inference with a more solid result.

References

Alcayaga, A., Wiener, M., & Hansen, E. G. (2019). Towards a framework of smart-circular systems: An integrative literature review. *Journal of Cleaner Production, 221*, 622–634. https://doi.org/10.1016/j.jclepro.2019.02.085.

Blondel, V. D., Guillaume, J.-L., Lambiotte, R., & Lefebvre, E. (2008). Fast unfolding of communities in large networks. *Journal of Statistical Mechanics: Theory and Experiment, 2008*(10). https://doi.org/10.1088/1742-5468/2008/10/p10008.

Hatchuel, A., & Weil, B. (2009). C-K design theory: An advanced formulation. *Research in Engineering Design, 19*(4), 181–192. https://doi.org/10.1007/s00163-008-0043-4.

Lee, J., Bagheri, B., & Kao, H.-A. (2015). A cyber-physical systems architecture for industry 4.0-based manufacturing systems. *Manufacturing Letters, 3*, 18–23. https://doi.org/10.1016/j.mfglet.2014.12.001.

Li, X., Chen, C.-H., Zheng, P., Wang, Z., Jiang, Z., & Jiang, Z. (2020). A knowledge graph-aided concept–knowledge approach for evolutionary smart product–service system development. *Journal of Mechanical Design, 142*(10). https://doi.org/10.1115/1.4046807.

Li, X., Jiang, Z., Guan, Y., Li, G., & Wang, F. (2019). Fostering the transfer of empirical engineering knowledge under technological paradigm shift: An experimental study in conceptual design. *Advanced Engineering Informatics, 41*. https://doi.org/10.1016/j.aei.2019.100927.

Li, X., Jiang, Z., Liu, L., & Song, B. (2018). A novel approach for analysing evolutional motivation of empirical engineering knowledge. *International Journal of Production Research, 56*(8). https://doi.org/10.1080/00207543.2017.1421785.

Li, X., Jiang, Z., Song, B., & Liu, L. (2017). Long-term knowledge evolution modeling for empirical engineering knowledge. *Advanced Engineering Informatics, 34*, 17–35. https://doi.org/10.1016/j.aei.2017.08.001.

Liu, L., Song, W., & Han, W. (2020). How sustainable is smart PSS? An integrated evaluation approach based on rough BWM and TODIM. *Advanced Engineering Informatics, 43*. https://doi.org/10.1016/j.aei.2020.101042.

Liu, B., Zhang, Y., Zhang, G., & Zheng, P. (2019). Edge-cloud orchestration driven industrial smart product-service systems solution design based on CPS and IIoT. *Advanced Engineering Informatics, 42*. https://doi.org/10.1016/j.aei.2019.100984.

Michelini, G., Moraes, R. N., Cunha, R. N., Costa, J. M. H., & Ometto, A. R. (2017). From linear to circular Economy: PSS conducting the transition. *Procedia CIRP, 64*, 2–6. https://doi.org/10.1016/j.procir.2017.03.012.

Shen, J., Erkoyuncu, J. A., Roy, R., & Wu, B. (2017). A framework for cost evaluation in product service system configuration. *International Journal of Production Research, 55*(20), 6120–6144. https://doi.org/10.1080/00207543.2017.1325528.

Taguchi, G. (1995). Quality engineering (Taguchi methods) for the development of electronic circuit technology. *IEEE Transactions on Reliability, 44*(2), 225–229. https://doi.org/10.1109/24.387375.

Wang, Z., Chen, C.-H., Zheng, P., Li, X., & Khoo, L. P. (2019a). A novel data-driven graph-based requirement elicitation framework in the smart product-service system context. *Advanced Engineering Informatics, 42*, 100983. https://doi.org/10.1016/j.aei.2019.100983.

Wang, Z., Chen, C. H., Zheng, P., Li, X., & Khoo, L. P. (2019b). A graph-based context-aware requirement elicitation approach in smart product-service systems. *International Journal of Production Research*. https://doi.org/10.1080/00207543.2019.1702227.

Westkämper, E., Alting, & Arndt. (2000). Life cycle management and assessment: Approaches and visions towards sustainable manufacturing (keynote paper). *CIRP Annals, 49*(2), 501–526. https://doi.org/10.1016/s0007-8506(07)63453-2.

Zhang, Y., Ren, S., Liu, Y., Sakao, T., & Huisingh, D. (2017). A framework for big data driven product lifecycle management. *Journal of Cleaner Production, 159*, 229–240. https://doi.org/10.1016/j.jclepro.2017.04.172.

Zheng, P., Chen, C.-H., & Shang, S. (2019). Towards an automatic engineering change management in smart product-service systems — a DSM-based learning approach. *Advanced Engineering Informatics, 39*, 203–213. https://doi.org/10.1016/j.aei.2019.01.002.

Zheng, P., Lin, T.-J., Chen, C.-H., & Xu, X. (2018a). A systematic design approach for service innovation of smart product-service systems. *Journal of Cleaner Production, 201*, 657–667. https://doi.org/10.1016/j.jclepro.2018.08.101.

Zheng, P., Lin, Y., Chen, C.-H., & Xu, X. (2018b). Smart, connected open architecture product: An IT-driven co-creation paradigm with lifecycle personalization concerns. *International Journal of Production Research, 57*(8), 2571–2584. https://doi.org/10.1080/00207543.2018.1530475.

CHAPTER 10

Conclusions and future perspectives

Contents

10.1 Conclusion	229
10.2 Future perspectives	232
10.2.1 Smart human-product-service-system	232
10.2.2 Cognitive interaction of smart PSS	232
10.2.3 Software-defined industrial digital servitization	232

10.1 Conclusion

From the elaborative analysis of Smart PSS in the previous chapters, one may find that it results from the convergence of cutting-edge digital technologies and today's ever-increasing transition of business model toward servitization, which is also known as digital servitization (Chapter 1).

As a typical type of digital servitization paradigm, Smart PSS emphasizes the development of smart, connected products, and their generated digitalized services, as the solution bundle to be implemented in various engineering lifecycle stages for value propositions. Hence, it follows the basic principles of PSS by addressing the complex product-service family design to meet individual customer needs. But also, it owns the digital capabilities of connect, intelligence and analytic, by leveraging the ICT, big data analytics and AI techniques, which largely distinguishes itself from the other PSS paradigms as described in Chapter 2.

Empowered by those digital capabilities, the technical, economical, and social benefits of Smart PSS were underlined in Chapter 3, which serves as the fundamental to show its significance. Meanwhile, the unique characteristics for Smart PSS development were further outlined in Chapter 3, including closed-loop self-adaptable design, new IT-enabled value co-creation, and design with context-awareness. They together compose the main aspects and challenges to be addressed in this book.

To deal with the closed-loop self-adaptable design process, Chapter 4 presented a generic design entropy theory, which represented the iteration

Smart Product-Service Systems
ISBN 978-0-323-85247-0
https://doi.org/10.1016/B978-0-323-85247-0.00013-X

© 2021 Elsevier Inc.
All rights reserved.

of forward design and inverse design process of Smart PSS as the dynamic change of information and entropy in a balanced system. Hence, one may evaluate the performance of the Smart PSS development by assessing the conversion rate of information in both short and long term.

To address the value co-creation issue, a hybrid crowd-sensing mechanism was brought up in Chapter 5, by leveraging massive user-generated data and product-sensed data in a human–machine intelligence integrated manner to support the concept or value generation process. Both the data/information fusion and incentive mechanisms were provided to guide the value co-creation process in a stepwise manner.

To overcome the challenge of design with context-awareness, the book authors follow the classic DIKW model, where context-aware Smart PSS development process is interpreted as a data-driven, knowledge-intensive activity. Hence, one fundamental issue is to offer a proper context-based modeling technique. Chapter 6 introduced a generic graph-based product-service-context configuration method, of which the interrelationship between various product and service components, and their corresponding context information can be depicted in a knowledge graph. Hence, in-context requirement elicitation, solution recommendation, and predictions can be performed by the graph-based querying process with advanced AI techniques. Meanwhile, enabled by the cutting-edge CPS, the other issue is to establish a DT which can represents the Smart PSS design and optimization process in a dynamic and visualized manner remotely, with high fidelity simulation, monitoring and control capabilities. Therefore, Chapter 7 proposed a generic trimodel-based DT reference model, which enables the ambient-based product-service family design in the early stage, and its own reconfiguration/optimization in the usage stage holistically.

To better depict the significance and prosperity of Smart PSS, except for the only case studies given in Chapters 4–7, the authors provided a comprehensive survey to summarize all the relevant industrial implementations to date in Chapter 8. Based on their adoption along various stages of the engineering lifecycle, they were further divided into design, manufacturing, distribution, usage, and end-of-life stages, respectively.

Lastly, Smart PSS shares the similar ultimate goal as PSS, which aims to enhance the customized solution development in a sustainable manner. Therefore, the sustainability concern of Smart PSS was investigated in Chapter 9 as well. Unlike other types of PSS only considering about the sustainability of physical product/components, the extended lifespan, better resource efficiency, and reversibility of the cyber and physical sustainability were holistically addressed in this chapter. A knowledge based standardized PDCA procedure was introduced based on the ISO14001:2015 to further guide the sustainable development of Smart PSS in an ever-evolving manner.

The managerial implications of this book encompass the challenges faced by today's industrial companies' transition toward digital servitization, especially their concerns on how to develop and manage their own smart, connected products and digitalized services in a cost-effective manner, regardless of the operation in a B2B or B2C manner. For industrial Smart PSS (B2B manner), the internal technology stack of a smart factory or manufacturing shop floor plays a dominant role, of which design with context-awareness and closed-loop iteration are more critical than the value co-creation process among different stakeholders. This is resulted from the fact that industrial autonomous and machine intelligence with no human intervention are to be achieved. Nevertheless, for customer-oriented Smart PSS (B2C manner), value co-creation process can be more significant or equally important among the three aspects if not, so as to meet the dynamic change of customer needs and to maintain their loyalty. Meanwhile, resource utilization efficiency is the key sustainable factor for industrial Smart PSS, since the internal operation, labor cost, and manufacturing resource allocation compose a larger amount of expenditure of the company to maintain its success. On the other hand, extended engineering lifespan is more important for the customer-oriented Smart PSS, since most of the revenue comes from the services offered in a pay-per-use manner. And in fact, there is still a long way for both parties to concern about the reversibility factor of sustainability, due to the relatively low value regenerated. Some other managerial implications can also be derived from this book, including:

- High performance, low-cost smart open architecture product, as the information container, should be able to change or upgrade readily in a software-defined manner to make high margin revenue. Hence, instead of chasing after the software capitals, companies should pay more attention to the "digital container" itself.
- The value co-creation profit model can be established among stakeholders by a data-driven revenue-based manner. Manufactures/service providers can offer more accurate monetary or other forms of incentives to the users/customers based on the value co-creation result in a data-driven manner.
- Since Smart PSS relies much on the electronic compositions of SCPs, reversibility and e-waste management in a smarter manner will play a more influential role.

10.2 Future perspectives

As an ever-maturing paradigm, some potential future research directions of Smart PSS are highlighted here to welcome more open discussions and in-depth research and development.

10.2.1 Smart human-product-service-system

To the authors' own definition, the smartness of Smart PSS is largely based on the smart, enabling technologies so as to make intelligent decision making. Hence, the level of smartness is judged by the digital capabilities in the 5C architecture. However, there can be another way of interpretation: that the smartness is evaluated by the acceptance and expectation of the human beings (e.g., user experience), especially for the customer-oriented Smart PSS. Hence, similar to the concept of human–cyber–physical system, a novel concept of smart human-product-service system can be the next stop, of which it not simply involves the value co-creation of human-in-the-loop, but also reflects the dominance of human's role in a result-oriented manner. It also matches well with the future tendency toward result-oriented Smart PSS, of which the evaluation of the users/customers toward the performance of the Smart PSS, other than the technical functionality matters more.

10.2.2 Cognitive interaction of smart PSS

From the scope of Smart PSS defined in Chapter 3, one can find that it involves three levels from micro to macro, including product-service level, system level, and ecosystem level. Although the graph-based approach described in Chapter 6 offers a promising manner to present the cognitive relationship and interactions among the complex product and service components in a context-based manner, major challenges still remain in the transition from perceptive interaction toward cognitive interaction, especially in the human–SCP interactions in an eco-system level. For instance, the cognitive understanding between the smart transportation system and the autonomous vehicle drivers. Semantics-based approaches and other forms of multimedia support can be a promising manner to be explored by leveraging the frontier AI techniques.

10.2.3 Software-defined industrial digital servitization

Ensured by the advanced digital capabilities (i.e., networking, digitalization, and smartness) of Industrial IoT, manufacturing companies can readily adopt the digital servitization business model to enable their productivity with high revenue. In this scenario, software-as-a-service plays a dominant role and often replaces the hardware components in a "software-defined X" manner, mainly in fourfold: (1) from physical asset perspectives, digitalized services are generated based on the smart, connected products to perform values in a more flexible manner; (2) from networking perspective, software-defined network together with microservice architecture enables the rapid, frequent, and reliable delivery of large, complex service applications; (3) from digitalization perspective, virtualized products or their digital twins in a software environment can offer types of digitalized services seamlessly (e.g., remote monitoring); and (4) from smartness perspectives, both the cutting-edge knowledge graph and deep learning techniques, as the industrial AI enable the smart functionalities in the digitalized services. To this end, other than the Smart PSS itself, the overall software-defined industrial digital servitization paradigm should be explored to understand its readiness and realization.

Nomenclature

4R Redesign, Remanufacturing, Reuse, and Recycle
5V Data Volume, Variety, Veracity, Velocity, and Value Data
AD Axiomatic Design
APP Applications
AR Augmented Reality
C−K Model Concept-Knowledge Model
CAD Computer-Aided Design
CE Circular Economy
CPS Cyber-physical System
CPU Central Processing Unit
DET Design Entropy Theory
DIKW Data, Information, Knowledge, Wisdom
DT Digital Twin
GPS Global Position System
ICM Information Conversion Map
ICT Information and Communication Technology
IoT Internet-of-Things
KE Kansei Engineering
KG Knowledge Graph
MC Mass Customization
MCS Mobile Crowd Sensing
ML/DL Machine Learning/Deep Learning
PDCA Plan-Do-Check-Adjust
PLM Product Lifecycle Management
PSB Product-service Bundles
PSS Product-service System
Q&A Question and Answer
QFD Quality Function Deployment
RA Requirement Attribute
RE Requirement Elicitation
RG Requirement Graph
RUL Remaining Useful Life
SCOAP Smart, Connected Open Architecture Product
SCP Smart, Connected Product
SE Solution Evaluation
SR Solution Recommendation
SSPSS Sustainable Smart Product-Service System
TRIZ Teorija Rezhenija Izobretatelskih Zadach
UGC User-generated Content

Index

'*Note*: Page numbers followed by "f" indicate figures and "t" indicate tables.'

A

Adaptable design, 88
Analytic capability, 28
Application programming interface (API), 86–87
Artificial intelligence (AI), 1–2
Augmented reality (AR), 54
Autonomous configuration with context awareness, 207

B

Blueprint design, 54
Brainstorming, 75–76
Building information model (BIM), 174
Built-in-flexibility design, 87–88
Business model, 28–32, 30t–31t
Business-to-business (B2B) model, 85–86, 231

C

C-K theory, 88
Closed-loop design, 41–42, 229–230
Cognitive interaction, 232
Concept-knowledge (C-K) model, 184–185
Configurable components (CCs), 91
Configuration toolkit, 88–89
Constraint entities, 160
Content organization, 4–6
Context-aware DT system
  case studies, 170–176
    establishing, 170–174, 172f
    in-context solution design, 174–176, 175f
Context-context relationship (CC), 130
Context-service relationship (CS), 130
Conversion
  ability, 63–64
  plan, 71–72
Crowd-sensing. *See* Hybrid intelligence

Customer-oriented product-service systems, 26
Cyber-physical system (CPS), 14–15, 160, 185–186, 206–207

D

Data collection, 103–106
  layer, 124
Data-driven circularity, 206
Data-driven platform-based process, 38
Data fusion, 103–106
Data-information-knowledge wisdom (DIKW) model, 28, 230
Data integration, 164
Data storage, 125
Design entropy theory
  case studies, 75–76, 75f, 77f–80f
  conversion plan, 71–72
  definition, 54–70
  design methodologies, 55t–57t
  fundamentals, 59–62, 60f
  information conversion map (ICM) tool, 72–74, 73f
  information entropy, 63
  innovative design entropy, 64–66
  iterative design entropy, 66–70, 66f–67f
  methodologies challenges, 53–54
  self-adaptable design process, 70–72, 70f
  smart design concept, 53–54
Design evaluation phase, 42
Design structure matrix (DSM), 154
Design with context-awareness, 43
Development toolkits, 90–91
Digitalization, 1–3
  e-services *vs.*, 26
  motivation, 3–4
  Smart product-service systems (PSS), 4–6
  vision, 3–4

238 Index

Digital platform, 32, 33t—35t
Digital servitization. *See* Digitalization
Digital technologies, 27—28, 27f
Digital twin-enabled servitization,
14—15
cyber-physical system (CPS)
vs., 26
design/optimization
DT-driven product, 154—155
engineering product family, 153—154
design structure matrix, 151—152
digital twin (DT), 152—153
information and communication
technologies (ICT), 152—153
modular design, 151—152
platform-based scalable design,
151—152
product life cycle management (PLM),
152—153
Digital twin-enhanced product family,
160—165, 161f
ambient-based product family design,
161—163, 162f
context-aware product family design,
163—165, 164f
digital twin-enabled servitization,
151—155
DT-driven product family optimization,
165—170, 166f
DT-driven reconfiguration, 167
DT-driven reverse design, 167—170,
169f
DT-enhanced product family design,
160—165
trimodel-based generic framework,
155—159

E

E-commerce platform, 119
Economic sustainability concerns, 208
Ecosystem level, 22
Edge-cloud computing, 1—2
Embedded open toolkit, 87—89
Engineering lifecycle implementations
application scenarios, 193, 194t—195t
design stage, 183—185

adoptions in, 184t
concept generation, 184—185
concept-knowledge (C-K) model,
184—185
knowledge graph (KG), 184—185
reference architecture model industry
4.0 (RAMI 4.0), 184—185
requirement management, 183
digitalization dimension, 181—182
distribution stage, 187—189
adoptions in, 187t
digitalization, 187—188
servitization, 187—188
end-of-life stage, 191—193
adoptions in, 192t
smart recycling, 192—193
smart reuse, 192
implementations, 182f
lifecycle dimension, 182
manufacturing stage, 185—187
adoptions in, 186t
service-oriented smart manufacturing
(SoSM) framework, 185
smart inspection, 187
smart production, 186—187
usage stage, 189—191
prognostic and health management
(PHM), 189—190
smart operation/maintenance,
189—190
smart reconfiguration, 191
Web of Science (WoS), 183
Environmental sustainability concerns, 207
E-services, 26

G

Global positioning system (GPS), 27
Graph-based product-service-context
modeling
application layer, 123—124, 132—133
data collection layer, 124
data storage, 125
graph construction layer, 123
graph model construction, 127—132,
127f
hypergraph construction, 130—132

requirement graph (RG), 128–130, 129f
heterogeneous data collection and storage, 125
hyperedge PSB-RA, 131
hyperedge PSB-US, 131
involved data, 125
knowledge management, 123–124
ontologies construction, 126–127, 126f
resource layer, 123
usage scenario (US), 130–131
user-generated data content (UGC), 125
Graph construction layer, 123

H

Heater malfunction (HM), 109–110
Heterogeneous data collection and storage, 125
Hybrid intelligence, 40–41
  case studies, 109–112
  data collection, 103–106
  fundamentals, 98–100, 98f
  generic framework, 100–102, 100f
    networked platform layer, 101
    physical resource layer, 100–101
    service application layer, 101–102
    service composition layer, 101
  incentive mechanism, 102–103, 102t
  information fusion, 103–105
  social sensors, 99–100
  value generation, cost-driven decision making, 107–108
Hyperedge PSB-RA, 131
Hyperedge PSB-US, 131
Hypergraph construction, 130–132
Hypergraph model
  case studies, 146–148
    hypergraph construction, 146–147
    PageRank, 147–148
    solution configuration, 147
  definition, 138
  offline training, 138–140
  online application, 141
  solution configuration based, 137–141
  unbiased hypergraph ranking approach, 138–140

I

Implementation architecture, 28
Incentive mechanism, 102–103, 102t
Industrial artificial intelligence, 14–15
Industrial Smart PSS, customer-oriented PSS vs., 26
Information and communication technologies (ICT), 1, 17
Information conversion map (ICM) tool, 72–74, 73f
Information entropy, 63
Information fusion, 103–105
Information theory, 62
Innovation, 2
Innovative design entropy, 64–66
Innovative design phase, 42
Intelligence capability, 28
Internet-based product-service systems (PSS), 16
Internet of Things (IoT), 1–2, 86–87
Internet-of-Things (IoT)-enabled product-service systems (PSS), 16
IT-driven product-service systems (PSS)
  internet-based product-service systems (PSS), 16
  IoT-enabled product-service systems (PSS), 16
  smart product-service systems (PSS), 17–18
Iterative design
  entropy, 66–70, 66f–67f
  phase, 42

K

Kansei Engineering (KE), 54
Knowledge graph (KG), 184–185
Knowledge management, 123–124

M

Mobile crowd-sensing and computing (MCSC), 99–100
Motivation, 3–4

N

Natural language processing (NLP), 185
Networked platform layer, 101

240   Index

Networking capability, 27
Node embeddings via SkipGram,
    135–136

O
Offline training, 138–140
Online application, 141
Ontologies, 157–158
    construction, 126–127, 126f
Open systems interconnection (OSI)
    model, 158
Original equipment manufacturers
    (OEMs), 92

P
PageRank, 147–148
Personalization, 164–165
Personalized solution, 119–120
Physical resource layer, 100–101
Plan-Do-Check-Adjust (PDCA)
        procedure, 210–218, 210f
    case studies, 218–225, 218f, 224t
    context awareness, 210–211, 211t
    DIKW, 211f
    knowledge evolvement (KE), 216–218
    requirement elicitation (RE), 212–213
    solution evaluation (SE), 215–216
    solution recommendation (SR),
        213–214
Product-context relationship (PC), 130
Product-dependent digital platform, 32
Product independent digital platform, 32
Product life-cycle management (PLM),
    26
Product-product relationship (PP), 128
Product-service family configuration
    context awareness, 120–121
    context-aware solution configuration,
        121–123
    e-commerce platform, 119
    graph-based product-service-context
        modeling, 123–133
    personalized solution, 119–120
    smart PSS, 118–123
Product-service systems (PSS), 204

business models, 10
classification, 11t–14t
definition, 9
IT-driven product-service systems
    (PSS), 14–18, 15f
service object-oriented system, 10
sustainability-oriented system, 10
Prognostic and health management
    (PHM), 175–176, 189–190

Q
Quality function deployment (QFD), 54

R
Radio-frequency identification (RFID),
    27
Reference architecture model industrie
    4.0 (RAMI 4.0), 158, 184–185
Requirement graph (RG), 128–130,
    129f
    context, 128
    context-context relationship (CC),
        130
    context-service relationship (CS), 130
    product-context relationship (PC), 130
    product-product relationship (PP), 128
    products components, 128
    services modules, 128
Requirement management
    case studies
        requirement elicitation, 144–145,
            145t
        requirement graph construction,
            141–144
        results discussion, 145–146
    graph model, 133–137
    node embeddings via SkipGram,
        135–136
    nodes prediction, 136–137
    requirement graph, 134–135
    sequence order via random walk, 135
Requirements analysis phase, 42
Research and development (R&D),
    85–86
Resource layer, 123

S

Self-adaptable design process, 70−72, 70f, 229−230
Sequence order via random walk, 135
Service application layer, 101−102
Service composition layer, 101
Service object-oriented system, 10
Service-oriented smart manufacturing (SoSM) framework, 185
Services modules, 128
Small and-medium-sized enterprises (SMEs), 92
Smart, connected open architecture product (SCOAP), 92−93
  value co-creation, service modeling for, 94−98
Smart, connected products (SCPs), 1
Smart design concept, 53−54
Smart human-product-service-system, 232
Smart inspection, 187
Smartness, 17
Smart production, 186−187
Smart product-service system (Smart PSS), 3
  basic notions, 21−26
  business aspect, 28−38
    business model, 28−32, 30t−31t
    digital platform, 32, 33t−35t
    value co-creation, 32−38, 36t−37t
  definition, 22, 23t−24t
  ecosystem level, 22
  fundamentals of, 38−44
    characteristics, 40−44, 41f
    design with context-awareness, 43
    development process, 38−40, 39t, 40f
    IT-driven value co-creation, 42
    sustainability, 44, 45t
  product-service level, 22
  scope, 22
  system level, 22
  system-of-systems level, 22
  technical aspect, 27−28
    analytic capability, 28
    digital technologies, 27−28, 27f
    implementation architecture, 28

intelligence capability, 28
    networking capability, 27
Smart travel assistant system (STA), 75
Smart water dispenser product (SWD), 109
Social sensors, 99−100
Sociotechnical ecosystem, 42
Software-defined industrial digital servitization, 232−233
Sustainable smart product-service systems, 10, 44, 45t
  autonomous configuration with context awareness, 207
  concept of, 205f
  cyber-physical resource reallocation, 206−207
  data-driven circularity, 206
  definition, 205−206
  economic sustainability concerns, 208
  environmental sustainability concerns, 207
  features, 206−207
  fundamentals of, 205−208
  Plan-Do-Check-Adjust (PDCA) procedure, 210−218, 210f
    case studies, 218−225, 218f, 224t
    context awareness, 210−211, 211t
    DIKW, 211f
    knowledge evolvement (KE), 216−218
    requirement elicitation (RE), 212−213
    solution evaluation (SE), 215−216
    solution recommendation (SR), 213−214
  product-service system (PSS), 204
  promoting directions for, 203−204
  social sustainability concerns, 208
  Sustainable Smart PSS (SSPS), 204
  systematic framework for, 208−209, 209f
  user-oriented long-lasting evolving, 207
  volume, velocity, variety, veracity, value (5V) data, 208−209
Sustainable Smart PSS (SSPS), 204
System level, 22
System-of-systems level, 22

T

Three-way incentive model, 111–112
Total dissolved solids (TDS), 109
Transition, digital servitization, 1–3
Trimodel-based DT reference model, 230
Trimodel-based generic framework, 155–159
  DT architecture featuring ambient information, 156–158, 156f
  in-context solution generation, 158–159, 159f

U

Unbiased hypergraph ranking approach, 138–140
Universal unique identifier (UUID), 100–101
Usage scenario (US), 130–131
User-centered design (UCD), 54
User-generated data content (UGC), 125

V

Value co-creation, 32–38, 36t–37t, 231
  adaptable design, 88
  built-in-flexibility design, 87–88
  C-K theory, 88
  design theory perspective, 87–88
  embedded open toolkit, 87–88
    IT-driven value co-creation toolkits, 88–92
    configurable components (CCs), 91
    configuration toolkit, 88–89
    development toolkits, 90–91
    embedded open toolkit, 89
    evolvement of, 90f
  open architecture product (OAP), 92–94
  perceptive mechanism, 38
  responsive mechanism, 38
  Smart, connected open architecture product (SCOAP), 94–98
    characteristics of, 96
    modular design, 96–97
    scalable design, 97–98
    two-stage design process, 95f
  two-stage based product design process, 87
Value-driven co-creation process, 38
Value generation, cost-driven decision making for, 107–108
Value-generation process, 40–41
Virtual prototyping, 86–87
Vision, 3–4
volume, velocity, variety, veracity, value (5V) data, 208–209

W

Web of Science (WoS), 183
Wireless sensor network (WSN), 27

Printed in the United States
by Baker & Taylor Publisher Services